THE CHEMISTRY OF URANIUM

TOPICS IN INORGANIC
AND GENERAL CHEMISTRY

A COLLECTION OF MONOGRAPHS EDITED BY

P. L. ROBINSON

Emeritus Professor of Chemistry in the University of Durham and the University of Newcastle upon Tyne

MONOGRAPH 13

Other titles in the collection:

THE CHEMISTRY OF URANIUM

INCLUDING ITS APPLICATIONS IN NUCLEAR TECHNOLOGY

BY

E. H. P. CORDFUNKE

Research Chemist,
Reactor Centrum Nederland,
Petten (The Netherlands)

ELSEVIER PUBLISHING COMPANY

AMSTERDAM / LONDON / NEW YORK

1969

ELSEVIER PUBLISHING COMPANY
335 JAN VAN GALENSTRAAT
P.O. BOX 211, AMSTERDAM, THE NETHERLANDS

ELSEVIER PUBLISHING CO. LTD.
BARKING, ESSEX, ENGLAND

AMERICAN ELSEVIER PUBLISHING COMPANY, INC.
52 VANDERBILT AVENUE
NEW YORK, N.Y. 10017

STANDARD BOOK NUMBER 444-40803-7

WITH 39 ILLUSTRATIONS AND 25 TABLES

PRINTED IN THE NETHERLANDS

To Ellie and Ine

Preface

Although uranium was discovered more than 150 years ago, interest in its properties dates only from the second world war. Because of its nuclear applications an enormous amount of work has had to be done since that time on the chemistry of the element, all in a relatively short time. This is clearly reflected in the considerable increase in the number of investigations published during the past twenty years.

The well-known monograph on the chemistry of uranium by Katz and Rabinowitch covered most of the work carried out before and during the second world war and shortly thereafter (up to 1950). A number of monographs on specialized topics have appeared since. Now the time seems ripe for a new book that gives a broad survey of the current chemistry and technology of uranium.

The present monograph is not exhaustive, but is intended as an introduction to the considerable amount of research now being done. A portion of the book deals with the technological aspects of uranium and its compounds: these are the fabrication of nuclear fuels and their behaviour, and the processing and reprocessing of these materials. In the opinion of the author it is difficult and not profitable to separate the chemistry and technology of the element. Their mutual interaction has resulted in its extensive application in nuclear technology.

The author is greatly indebted to Professor P. L. Robinson for critically reading the draft of each chapter and for the many helpful suggestions which he made.

The author wishes to thank some of his colleagues at the Reactor Centrum Nederland, particularly Mr. G. Prins, Dr. B. O. Loopstra, and Mr. W. Ouweltjes, for reading all, or part, of the manuscript and for their suggestions, and Mr. H. Kooistra for the careful drawing of the figures. The preparation of the subject index, done by Mr. G. Prins, is gratefully acknowledged.

July 1969   E. H. P. Cordfunke

Contents

LIST OF ABBREVIATIONS USED FOR
SCIENTIFIC REPORTS

AECD AEC-tr.	United States Atomic Energy Commission, Division of Technical Information Extension.
AERE-R -M	Atomic Energy Research Establishment, Harwell, Didcot, Berkshire, United Kingdom.
ANL	Argonne National Laboratory, Lemont, Ill., U.S.A.
BMI	Battelle Memorial Institute, Columbus, Ohio, U.S.A.
BM-RJ	Bureau of Mines, U.S.A.
BNWL	Battelle Northwest, Richland, Wash., U.S.A.
CEA-R	Commissariat à l'Energie Atomique, Paris, France.
DP	E. I. du Pont de Nemours Co., Wilmington, Del., U.S.A.
Eur	European Atomic Energy Community.
Euraec	United States - Euratom Joint Research and Development Program.
GAT	Goodyear Atomic Corp., Portsmouth, Ohio, U.S.A.
GEAP	General Electric Co.; various locations.
IGR-TN/CA	United Kingdom Atomic Energy Authority; various locations.
ISC-	Ames Lab., Ames, Iowa, U.S.A.
KAPL-	Knolls Atomic Power Laboratory, Schenectady, N.Y., U.S.A.
KFK	Gesellschaft für Kernforschung mbH., Germany.
KR	Institutt for Atomenergie, Kjeller, Norway.
LA	Los Alamos Scientific Lab., N.Mex., U.S.A.
MCW	Mallinckrodt Chemical Works, St.Louis, Mo., U.S.A.

MDDC   United States Atomic Energy Commission,
         Manhattan District, Oak Ridge, Tenn., U.S.A.

NAA-SR   Atomics International, Division of North American
         Aviation, Inc., Canoga Park, Calif., U.S.A.

NLCO   National Lead Co. of Ohio, Cincinnati, Ohio,
         U.S.A.

NYO   New York Operations Office, New York, N.Y.,
         U.S.A.

ORNL   Oak Ridge National Laboratory, Tenn., U.S.A.

TID   Office of Technical Information Extension,
         U.S. Atomic Energy Commission, Oak Ridge,
         Tenn., U.S.A.

CHAPTER 1

Introduction

History

The element uranium was discovered in 1789 in the pitchblende ores of Saxony by M.H. Klaproth (1743–1817), a notable analytical chemist and, after 1810, professor at the university of Berlin[1]. The discovery of uranium was the final step in attempts that had been going on for a long time to classify the mineral pitchblende, found in Joachimsthal (Bohemia) and also in Johanngeorgenstadt (Saxony)[2]. On treating pitchblende with nitric acid, Klaproth obtained a yellow solution, from which a yellow precipitate could be isolated when "potash" was added. He considered this precipitate to be the oxide of a new element and indeed secured a substance which had a metallic appearance when it was heated with carbon at a high temperature. Klaproth believed this to be the metal itself, but we now know that it was in fact a uranium oxide.

Klaproth described his discovery on the 24th of September 1789 in a lecture given before the Royal Prussian Academy of Science, entitled *Über den Uranit, ein neues Halbmetall*[3]. He thus named the new element "Uranit", after the planet Uranus, discovered some years before (1781). A year later Klaproth changed the name to uranium: "Den Namen des neuen Metalls, Uranit, habe ich rektifiziert und, den Regeln der Analogie gemäß, in Uranium verwandelt"[4].

By 1789 Klaproth had already isolated the nitrate, sulphate, acetate and formate of uranium as well as potassium and sodium uranates. In many publications he described the preparation and properties of these compounds. But till his death in 1817 he remained convinced that he had isolated the element.

References p. 5

The discovery of uranium in pitchblende and the preparation of some of its salts certainly caused a sensation. Indeed, in 1790, J. G. Leonhard proposed naming the new element "klaprothium" and J. B. Richter dealt with its properties and those of its compounds in his book *Über die neuen Gegenstände der Chemie* that appeared in Breslau in 1791.

Berzelius, who studied the material "uranit" soon after its discovery, suspected that it was not the element, but was unable to reduce it further by the use of potassium. It was more than half a century after Klaproth's discovery that the French chemist E. M. Péligot proved with certainty that "uranit" was not the element itself, but an oxide. This he did by passing chlorine over a heated mixture of "uranit" and carbon and so produced besides uranium chloride, a mixture of CO and CO_2. Péligot was also the first to prepare the metal. In 1841, he heated the anhydrous, sublimed uranium tetrachloride with potassium in a platinum crucible; after the product had been cooled, potassium chloride was leached out with water. The black residue contained pieces of uranium with a silvery lustre. The density of the metal was not determined.

Later, in 1856, Péligot prepared a purer product by melting UCl_4 with sodium in porcelain in the absence of air; the density of this material was 18.4, which is reasonably close to the value 19.05 now accepted for the pure metal. He also described some of the more important properties of uranium; for instance, its rapid oxidation, especially at higher temperatures. He also determined its atomic weight, for which he found the value 120. However, D. I. Mendeleev in 1869 was unable to accept this since it brought uranium between silver (108) and iodine (127) in his periodic table. This place was not in accordance with its properties and he therefore doubled the value. As a result he obtained a good periodicity in the properties of uranium and the analogous elements tungsten, molybdenum and chromium. But—as he writes[5]—"Die vorgeschlagene Abänderung im Atomgewicht des Urans läßt die Natur seiner Verbindungen von einem veränderten Gesichtspunkte aus betrachten und fordert daher zu neuen Untersuchungen über den Grad der Ähnlichkeit mit Cr, Mo, W auf".

Indeed, several investigations into the properties of uranium were made in the years following. In particular those by C. Zimmermann[6] confirmed that the atomic weight of uranium had been correctly taken to be about 240.

Determination of the vapour density of UCl_4 and UBr_4, according to the method of Victor Meyer, yielded the value 242.4 for the atomic weight of uranium. At that time, interest in the properties of the element was purely academic. Its compounds were of little economic importance and used only on a very small scale, mainly for colouring glass and porcelain. Uranium metal itself was tried at the end of the 19th century as a substitute for tungsten in tool steels for war applications[2]. But none of these applications involved quantities of any importance and the deposits of uranium ores at Joachimsthal (Austria) were mined mainly to satisfy the demand for thorium.

This was the position until the discovery of radioactivity by Becquerel (1896) and later that of the presence of radium in uranium ores (1898). When the demand for radium suddenly increased, the deposits of uranium ores at Joachimsthal and those in the Belgian Congo and in Canada were mined for their radium content. But until 1942 no mining operations were carried out primarily for uranium production.

With the advent of the nuclear age, brought about by the almost simultaneous discovery of nuclear fission (1938/39) and the outbreak of the second world war, a new interest in the element uranium and its compounds was born and has grown rapidly since that time.

During and after the second world war, much new information on the chemistry of uranium had been gathered in the USA as part of the Manhattan project. It was only after the first "Conference on the peaceful uses of atomic energy", held in Geneva in 1955, that much of this knowledge became generally available. The nuclear reactor now acts as a focus for practically all the current research in uranium chemistry, and this is the field with which the present book is concerned.

References p. 5

Nuclear properties of uranium

Natural uranium is a mixture of three isotopes, ^{238}U, ^{235}U and ^{234}U (Table 1). Other short-lived isotopes of the element have been made by nuclear reactions.

TABLE 1

NATURALLY OCCURRING ISOTOPES OF URANIUM

Isotope, mass	Atomic percentage[7]	Half-life, years	Ref.
238	99.276 ±0.0005	$4.51 \cdot 10^9$	8
235	0.718 ±0.0005	$7.09 \cdot 10^8$	9
234	0.0056±0.0001	$2.35 \cdot 10^5$	8

Dempster in 1935 was the first to recognize indications of the presence of ^{235}U by mass-spectrometry[10]; he estimated its content as < 1% in natural uranium. Later, Nier determined the abundance ratio 238/235 = 139±1. A later determination made on uranium ores from different sources[10] at the Columbia University yielded for the 238/235 ratio the value 138.0±0.3. This was confirmed by Greene et al.[11], who found the value 137.96±1.4.

The atomic weight of uranium has been revised several times in the history of the element. The value 238.14 was adopted in 1928 by the International Committee for Atomic Weights. This value was certainly too high, and newer determinations by Hönigschmidt and Witner[12] in 1936 yielded the value 238.07. The revision of the atomic weights in 1962, now based on ^{12}C, has brought its atomic weight to the value 238.03, this being the value now internationally accepted[13].

The isotopes ^{238}U and ^{234}U belong to the same radioactive decay series. This is the $(4n+2)$ series; of this series ^{238}U is the parent, and from it ^{234}U is formed by α- and β-decay. The ratio of the two isotopes is thus constant and equal to the ratio of their half lives. Most of the radioactivity of uranium ores comes from escaping ^{222}Rn and its short-lived decay products. The half-life of ^{222}Rn is 3.82 days, and the removal of radon from uranium ores thus disposes of most of their α- and γ-activity. Its absence is, however, only for a short time, since after a period of about ten half-lives the radon concentration has returned to the equilibrium value and its daughters will have reappeared. The precursor of ^{222}Rn is ^{226}Ra which has a half-life of $1.62 \cdot 10^3$ years and removal of radium effectively prevents the formation of ^{222}Rn and thus eliminates most of the α- and γ-activity. Normally radium is separated from uranium at the ore-leach stage. It constitutes a health hazard in the uranium ore treatment, since it is highly radioactive and tends to become fixed in the bones.

The β-activity of uranium ore is chiefly due to ^{234}Pa with a half-life of 1.14 min. This element is also removed during uranium purification together with its precursor ^{234}Th (half life = 24.1 days), but grows fairly rapidly and the β-activity of purified uranium returns to its original value within a year.

The α-activity of purified uranium is due to the naturally occurring isotopes of the element.

REFERENCES

1 G. E. DANN, *Martin Heinrich Klaproth*, Berlin (1958).
2 F. KIRCHHEIMER, *Das Uran und seine Geschichte*, Stuttgart (1963).
3 M. H. KLAPROTH, *Chem. Ann.* (1789), Bd. 2, 387. The original text of the lecture appeared in 1792 in French in: *Mém. Acad. Roy. Sci. Berlin* (1792), p. 160–174.
4 M. H. KLAPROTH, *Chem. Ann.* (1790), Bd. 1, 292.
5 D. MENDELEEV, *Ann. Chem. u. Pharm.* VIII, Supplementbd., 2. Heft (1872), 182.
6 C. ZIMMERMANN, *Ann. Chem.*, 232 (1886) 299.
7 F. A. WHITE, T. L. COLLINS AND F. M. ROURKE, *Phys. Rev.*, 101 (1956) 1786.
8 E. H. FLEMING Jr., A. GHIORSO AND B. CUNNINGHAM, *Phys. Rev.*, 88 (1952) 642.
9 Ph. O. BANKS and L.T. SILVER, *J. Geophys. Rev.*, 71 (1966) 4037.
10 J. J. KATZ and E. RABINOWITCH, *The Chemistry of Uranium*, Part I. McGraw-Hill Book Company, Inc., New York (1951).
11 R. E. GREENE, C. A. KIENBERGER AND A. B. MESERVEY, K-1201 (1955).
12 O. HÖNIGSCHMID and F. WITTNER, *Z. Anorg. Allgem. Chem.*, 226 (1936) 289.
13 A. E. CAMERON AND E. WICHERS, *J. Am. Chem. Soc.*, 84 (1962) 4195.

CHAPTER 2

The Extractive Metallurgy of Uranium

Uranium ores

The element uranium is widely distributed in the Earth's crust and in the hydrosphere. The uranium concentration in sea water (3.34 μg/l) appears to be remarkably constant[1]. The Earth's crust contains about 0.0004% uranium which is more than, for instance, the gold, silver and mercury contents and about equal to that of tin.

Uranium is never found in an elemental state, but always in chemical combination with other elements, with which it forms about 150 known minerals[2]. These minerals occur as separate grains or in veins in different types of rock; only a few of these are of commercial importance at present. The uranium ores, that is those minerals that can be mined and treated as an economically viable source of the element, usually have a uranium content up to 0.1%[3]. Uranium generally occurs in high-silica, acid rocks, such as granite. The basic rocks (basalts) contain a significantly lower uranium percentage than the average of the Earth's crust. Normally the sedimentary rocks contain negligible amounts of uranium, although there are important exceptions, such as the carnotite-bearing sandstones of Colorado.

The uranium minerals are broadly divided on the basis of occurence into two classes, primary and secondary[4]. The primary minerals, with the exception of pitchblende, appear in pegmatites and are presumably formed during terminal stages in the solidification of an acid magmatic intrusion. These primary minerals are readily modified by weathering or hydrothermal action into a wide range of secondary minerals. The alteration involves the oxidation of uranium; so that most of these secondary minerals, such as the phosphates, sulphates and silicates, are brightly coloured: for instance, the mineral carnotite ($K_2O \cdot 2UO_3 \cdot V_2O_5 \cdot nH_2O$), found in Utah and Colorado, is canary-yellow.

Of uranium minerals, uraninite is reported as being the most widely distributed. It is a primary mineral, found only in pegmatites, and is dark in colour. The composition is essentially that of uranium dioxide, UO_2, although some further oxidation of the uranium frequently occurs; the degree of oxidation depends on the local geological conditions. Another oxide mineral is pitchblende, a microcrystalline, nonpegmatitic ore, with the composition U_3O_8. It is often mistakenly described as a variety of uraninite, the terms uraninite and pitchblende being used as synonyms, which is obviously incorrect since they differ in composition and property. Pitchblende usually contains no thorium and only traces of the rare earths, in contrast to uraninite. Pitchblende is found in veins where it has been deposited from hydrothermal solutions.

The other primary uranium minerals[5] are complex oxides of which brannerite and davidite (Table 2) are the most significant. Davidite is the ore characteristic of the high-temperature deposit at Radium Hill (Australia) and brannerite is the mineral which

TABLE 2

SOME IMPORTANT URANIUM ORES WITH THEIR COMPOSITION

Ore	Composition	Uranium content, %	Typical occurrence
Uraninite	UO_{2+x}	45–85	Blind River (Canada)
Brannerite	$(U, Ca, Fe, Th, Y)_3 Ti_5 O_{16}$	40	Blind River (Canada)
Davidite	$(U, Fe, Ce) (Ti, Fe, V, Cr)_3 (O, OH)_7$	$\sim$10	Rum Jungle (Australia)
Autunite	$CaO \cdot 2\,UO_3 \cdot P_2O_5 \cdot 8\text{–}12\,H_2O$	50	France
Carnotite	$K_2O \cdot 2\,UO_3 \cdot V_2O_5 \cdot 3\,H_2O$		Colorado, Utah (USA)
Uranophane	$Ca(UO_2)_2 Si_2O_7 \cdot 6\,H_2O$	57	Congo
Coffinite	$U(SiO_4)_{1-x}OH_{4x}$		Colorado

generally exceeds uraninite in quantity in the uranium occurrence at Blind River in Canada.

Of the secondary minerals, autunite, the hydrated calcium uranyl phosphate found in France, is one of the commonest uranium minerals. Others are carnotite and tucholite.

Before the advent of atomic energy, uranium was a little con- sidered by-product of radium recovery at Great Bear Lake (Canada), Shinkolobwe (Congo) and Joachimsthal (Czechoslo- wakia). Only small amounts of uranium were actually separated and the main use for its compounds was in the colouring of glass and ceramics. But the atomic energy project changed all this and uranium was the subject of an intense prospecting effort in the years following 1946. This failed, however, to locate many large bodies of high-grade ore, occurrences of which are now known to be very rare. Hence most of the world's uranium is extracted from low-grade ores with up to 0.1% uranium. An exception is the even lower-grade ores (0.02%) in South Africa, in which uranium is intimately associated with gold and is produced as a by-product of the gold extraction industry. At present, the bulk of the uranium, about 40%, comes from the low-grade, but extensive deposits in the Blind River areas of Canada and from the gold-bearing ores of Witwatersrand in South Africa.

Up to now the largest quantities available were those discovered in Canada (Blind River) in 1953. In the Blind River area, the main uranium-bearing minerals are uraninite and brannerite; these have an average rating of 0.12% U_3O_8. Besides uraninite, the Blind River conglomerates contain about 0.05% ThO_2 and 0.025% rare-earth oxides. In recent years a substantial amount of thorium has been produced as a by-product from the working-up of this deposit.

The deposits in Canada belong to three types, namely conglo- meratic, veins and pegmatitic. The bulk of Canada's uranium and thorium reserves are in the conglomeratic ores near Elliot Lake, Ontario; they contain 93% of the reserves and have an average grade of 0.12% U_3O_8 and 0.05% ThO_2. Reserves in pitchblende- bearing vein-type deposits in the Beaverlodge Lake area of

Saskatchewan are less important; they comprise 6% of the total Canadian reserves[6]. It is rarely possible to develop ore reserves in vein-type deposits far in advance of the actual mining operations. The least important are the pegmatitic deposits (Bancroft, Ontario); they make up about one percent.

In the United States of America uranium deposits, although they occur in widely different geological conditions, are mainly confined to the Colorado Plateau type of deposits, and, less frequently, to the vein deposits. About 95% of the ore reserves (0.23% U_3O_8) are found in the Colorado Plateau.

In Europe, the most important uranium reserves at present known are in France. In that country most of the uranium occurs in vein deposits in granitic rocks; the average rating is 0.15% U_3O_8.

The development of nuclear energy depends very much on considerable amounts of uranium being available at reasonable prices. This implies that uranium reserves must be sufficiently large and that the uranium content of the ores constituting them must not be too low. Uranium resources have been divided into three price classes: $5–10/lb U_3O_8, $10–15/lb U_3O_8 and $15–30/lb U_3O_8. At present only those in the $5–10 range are of economic importance, and, as has been seen already, they generally have an average uranium content of 0.1%. Lower-grade ore cannot be treated economically for the recovery of its uranium content in an existing plant and the material, under present economic conditions, would not justify the erection of a new plant.

Uranium ores are usually found in discrete deposits of limited extent. Estimates of world potential can only be made on the basis of geological knowledge of the mode of occurrence of the known abnormal concentration of uranium (ore bodies) and on the chances of similar bodies being found elsewhere. Since only a small fraction of the Earth's surface is known geologically in detail, any estimate must be very tentative (Table 3). The largest areas currently known are in Canada, the United States, and Australia. The production of uranium in Canada increased in a spectacular way between 1955 and 1959, during the period rising from 1000 to 15,900 short tons U_3O_8 per year. Since 1959, however, it has declined to the 1966 level of 3900 ton U_3O_8 per year, and this rate of production will be maintained until 1970[8].

TABLE 3

URANIUM RESERVES [7](IN SHORT TONS* OF U_3O_8)

	Year ending		
	1958	1961	1966
Canada	414,577	277,968	210,000
U.S.A.	220,750	175,000	200,000
Republic of South Africa	370,000	150,000	205,000
Other countries	60,000	80,000	100,000
	1065,327	682,968	715,000

*One short ton U_3O_8 = 0.77 metric ton uranium metal.

The uranium required to fuel the future nuclear capacity could be calculated from a knowledge of the fuel requirements of the different types of reactors. These requirements vary from type to type, and, unfortunately, it is impossible to predict with any degree of certainty in what proportions the various types will be installed in the next 20 years. From reasonable assumptions, the Atomic Energy Commission of the United States has estimated in 1966 that the annual free-world requirement will rise from 12,000 short tons of U_3O_8 in 1970 to 65,000 short tons of U_3O_8 in 1980. It is evident, although forecasts of this nature require frequent revision, that the demand for uranium in 1975–1980 will be such that new sources will have to be found. To meet long-term requirements not only must new deposits be found, but also whole new fields of occurrence. Prospecting is very active now, especially in the United States and in Canada, so that large sources may still be expected to be found. Moreover, today's exploration efforts are supported by a much better knowledge of uranium geology than those made in 1950.

Since 1966 exploration in Canada has revived; most activity has been centered in Elliot Lake, Ontario, and Beaverlodge, Saskatchewan. Many companies have been acquiring land and are planning exploration programmes in the near future. At least one new area, near Agnew Lake, Ontario, has a sufficiently large tonnage to warrent underground development.

Geologically, there is every indication that experience in respect to uranium will follow that for other mineral commodities and, in response to market demand, continual exploratory effort will result in an ever-increasing expansion of reserves. A further point of importance is that cheaper methods of mining and processing will reduce the costs of the uranium recovery from the already discovered low-grade ores so that there is a possibility of increasing resources in the lower-priced range. In this connection it is interesting that the oceans contain considerable amounts of uranium and that the feasibility of a process for the recovery of uranium from sea water has been recently explored in some detail[9].

Concentration and extraction of uranium ores

The treatment of a variety of uranium ores of various grades and compositions has been adequately described[10,11] and only the general principles of the processes need to be discussed here.

Preconcentration

The first step in processing the ores, before the uranium is brought into solution by leaching, is to reduce the material to a small size by grinding. Much effort has been expended on preconcentration of the ores in order to reduce as far as possible the concentrate which has to be subjected to the expensive leaching process. This is of especially great importance in the processing of low-grade ores[8]. An example of such process is seen at the Radium Hill plant in Australia where the ore davidite, containing 0.6% uranium, is up-graded to a content of about 3% by physical methods of concentration.

Industrially, many different concentration techniques are employed, depending on the nature of the mineralization concerned.

Radiometric sorting is a relative new process which has rapidly become widespread. It is based on the radioactivity of small quantities or pieces of ore. The apparatus employs the measurement of both the radioactivity and the mass of the falling ore particles. The relevant ore pieces are then sorted, for instance by the rejection of ores containing less than 0.1% U_3O_8, by mechanical means into high-grade material and waste.

Gravimetric concentration has not been found to be widely applicable. It is mainly used in a preliminary way to separate the ore into the various fractions of different mineral composition. Flotation, however, has been used successfully, for instance to separate ore into the carbonate-rich and silicate-rich fractions for separate alkaline and acid leaching. At Radium Hill in Australia, flotation is used to clean the davidite concentrates obtained by heavy medium separation. Of the possible surface-active agents, sodium oleate was found to give the best results. As an example, the grade of a feed was 7.2 lb U_3O_8 per ton, whereas the grade of the concentrate was 18.3 lb U_3O_8 with a recovery of 85%.

Leaching

After preconcentration, the ores are leached to bring uranium into solution. This may be done either by treatment with sulphuric acid or with a sodium carbonate/sodium bicarbonate solution. Alkaline leaching is used only for materials, such as carbonates, which would consume a wasteful amount of acid. Its chief advantage is the relatively non-corrosive properties of the solutions employed and the fact that few impurities are dissolved along with the uranium. Air is almost universally used as the oxidant and elevated temperatures are essential. The use of the carbonate-leaching process for uranium extraction had its origin in the treatment of the carnotite deposits of the Colorado Plateau with vanadium recovery as the primary objective. When the uranium content of these ores became the most valuable constituents, the existing plants and processes were optimized for the uranium recovery.

In acid leaching, the ores are brought into contact with sulphuric acid of such a concentration that the pH of the final liquor is

1.5–1.8. Uranium ores with uranium in the tetravalent oxidation state, in particular the primary uranium minerals such as pitchblende, uraninite and tucholite, are insoluble in dilute sulphuric acid. To overcome the difficulty, the leaching is done in the presence of an oxidizing agent; for this MnO_2 or sodium chlorate are commonly used. When the oxidant is MnO_2, the requisite quantity is about 10 lb/ton of ore for most ores. The manganese is recovered after the extraction.

Leaching is the fundamental operation in the processing of uranium ores. It has a considerable influence on both the technological and economic aspects of the whole extraction. For these reasons it has attracted a great deal of attention and several improvements have been successfully applied, such as leaching at a controlled pH and at a controlled redox potential. In the latter case the potential between a platinum and a calomel electrode inserted in the digesting ore mixture is controlled at a value of 0.60 V; iron is then present in the Fe(III) form, indicating oxidizing conditions. Solutions at this potential usually yield satisfactory extractions from pitchblende during a 16 to 24 hour contact time. More aggressive leaching is sometimes needed, for instance in the dissolution of brannerite where the free acid concentration employed is about 5%. Heating during leaching is also employed; it is merely a matter of economics whether heat or an excess of acid is to be preferred. As the cost of acid is the major expense in the processing costs, economical use of the reagent is important. At Radium Hill in Australia leaching had to be carried out at the boiling point with strong sulphuric acid owing to the refractory nature of the ore (davidite).

Filtration is widely used for the separation of leach liquors from solid residues. The pregnant solution from the leaching, after clarification, contains about 0.5 g/1 uranium. For the purification and concentration of the dilute solutions the ion exchange method is commonly used. In this, the uranium is absorbed from the solution as an anionic complex $UO_2(SO_4)_2^{3-}$ or $UO_2(SO)_3^{4-}$ from the equilibrium:

$$UO_2^{2+} + n\ SO_4^{2-} \rightleftarrows UO_2(SO_4)_n^{(2n-2)-}$$

References p. 20

and the method has been developed to a high degree of efficiency. Anion exchange for the recovery of uranium from carbonate solutions is very similar in principle to that for the recovery from sulphuric acid solution. In it the $UO_2(CO_3)_3^{4-}$ ion is absorbed from the solution. The ion exchange process for the acid-leaching solution is done in two or three columns, connected in series, through which the pregnant solution is passed. The uranium loading is very sensitive to the pH of the solution. It falls from 110 g U_3O_8 per litre of resin at pH = 2.0 to 30 g U_3O_8 per litre of resin at pH = 0.5. In practise, the maximum pH used is about 2.0, because, at a higher pH, precipitation of uranium takes place. Impurities dissolved during the leaching of the ore, for instance manganese(II), calcium, magnesium, aluminium, copper and vanadium(IV) are not absorbed, but iron(III), vanadium(V) and molybdate, arsenate and phosphate are. They can be removed by suitable washing, the molybdate for instance with caustic soda solution. The adsorption of Fe(III) decreases sharply as the pH falls and is nearly zero at pH = 1.4.

The clarification of leach liquors is one of the more difficult and expensive steps in the processing of some uranium ores. When this trouble is serious the ion exchange process is done without complete clarification of the leach liquor, provided that the solids can be kept from mechanically interfering with the operation of the resin bed.

In the so-called *resin-in-pulp process*, used in some mills in Colorado, large granules of resin in wire baskets are oscillated vertically in a horizontal stream of the incompletely clarified leach liquor. The leaching is done in a two-stage or multi-stage counter current plant at 60 °C.

Uranium is removed from the resin by elution with solutions containing high concentrations of chloride or nitrate ions. Eluting solutions have either the composition $1M$ $NH_4NO_3 + 0.2$ N HNO_3 or $1M$ NaCl $+ 0.2$ N HCl. The high affinity of these ions for the resin and their high concentrations cause displacement of uranium from the resin into the solution, in which it reaches a concentration of 10–20 g/l. From this solution uranium is recovered economically by precipitation. When nitrate eluants are used, the precipitation is done with ammonia, but chloride eluants need caustic soda to obtain the sodium diuranate.

Purification

The next step in the process is the purification of the precipitated concentrates. The techniques of purification of the crude precipitates from the mines were largely developed in the USA. Originally, ether extraction had been used successfully, but was abandoned in favour of extraction with tributyl phosphate (TBP). This is done by dissolving the precipitates in nitric acid and treating the solution by solvent extraction with TBP dissolved in kerosene or *n*-hexane. The concentration of TBP in these solvents is between 25 and 40%.

The extraction of uranium is based on the formation of a complex compound between uranyl nitrate and TBP, according to the equation:

$$UO_{2(aq)}^{2+} + 2\,NO_{3(aq)}^{-} + 2\,TBP_{(org)} \rightleftarrows UO_2(NO_3)_2 \cdot 2\,TBP_{(org)}$$

The neutral complex contains no water and is soluble only in the organic phase. Uranyl nitrate is distributed over both phases. The

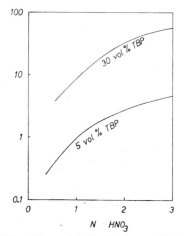

Fig. 1. Effect of nitric acid on the distribution coefficient of uranium; the solvent is tributyl phosphate (TBP) in kerosene at room temperature.

distribution coefficient $D = [UO_2(NO_3)_2]_{org}/[UO_2(NO_3)_2]_{aq}$ is influenced by the presence of nitric acid and nitrate ions. Figure 1 illustrates the effect of nitric acid concentration on the distribution coefficient. The addition of nitrates has a similar influence. These principles form the basis of industrial practice.

The commercial concentrates are dissolved in nitric acid to give a solution of about 300 g/l of uranium in 2–3 N free nitric acid. When sulphate anion is present it must be complexed by the addition of iron or aluminium. The solvent extraction is done in a three-step, counter-current process. Uranium is transferred to the solvent phase in an extraction section. The small amounts of impurities in the organic phase are removed in the scrub section. Finally, the uranium is recovered by back-extraction in a third section on washing the organic phase with very dilute nitric acid at a temperature of about 60 °C in order to depress the uranium distribution coefficient. The concentrated solution of uranyl nitrate thus obtained is then filtered through kieselguhr. The extracted organic phase is recycled after first being washed with water in 1 N nitric acid.

Production of uranium compounds

The product resulting from the solvent extraction process (see Fig. 2) is a clear solution of almost pure uranyl nitrate. It forms the starting material in the industrial production of a number of important uranium compounds[11,12]. For this, the first step is to prepare the oxide UO_3 ("orange oxide") from the concentrated solution. Two different methods are used to convert the purified uranyl nitrate solution into UO_3. In government owned plants of the United States, the solution is concentrated by evaporation, oxides of nitrogen being evolved, as the temperature continuously rises to about 450 °C. At this point the denitration is completed.

In Canada and in France, UO_3 is produced by decomposing a precipitate of ammonium diuranate (so called ADU), obtained by adding ammonia or urea to the nitrate solution. The UO_3 produced is subsequently reduced to the oxide UO_2 either with hydrogen or with cracked ammonia (a by-product from the decomposition of ADU). The brown coloured uranium dioxide, UO_2, which is of technological importance as a nuclear fuel, cannot easily be reduced directly to uranium metal. For this reason it is first hydrofluorinated

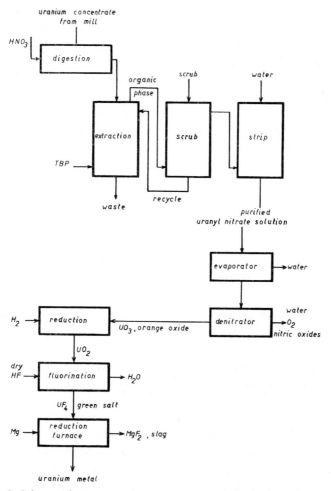

Fig. 2. Scheme of ore concentrate treatment and fabrication of uranium compounds.

to give uranium tetrafluoride UF_4 ("green salt"):

$$UO_3 + H_2 \xrightarrow{650\,°C} UO_2 + H_2O$$
$$UO_2 + 4HF \xrightleftharpoons{500\,°C} UF_4 + 2\,H_2O$$

The hydrofluorination of UO_2 is generally done in a moving bed and is an important chemical reaction, since the compound UF_4 is a key intermediate for the production of uranium metal and uranium hexafluoride, both important in nuclear technology. However, the large-scale production of UF_4 is a difficult operation because the rate and the extent of the reaction is dependent on a number of process variables, such as the time and the temperature of the reaction. Besides this, the physical properties of the UO_2 used (specific surface area as determined by the reduction temperature) has important effects.

Reduction of UF_4 to uranium metal is effected either with magnesium or calcium; proper operation produces uranium metal of commercial purity.

$$UF_4 + 2Mg \rightarrow U + 2\,MgF_2 + \;\;82\,kcal$$
$$UF_4 + 2\,Ca \rightarrow U + 2\,CaF_2 + 134\,kcal$$

The heat made available by the reaction must be sufficient to melt the uranium and render the slag sufficiently fluid; only under these conditions can the dispersed particles of the newly produced uranium flow together and coalesce after rejecting the slag. Reduction with calcium generates enough heat to melt the slag. But the heat generated during reduction with magnesium is not of itself sufficient to properly fuse the products. The necessary additional heat is provided by external heating of the reaction vessels. Notwithstanding this disadvantage, magnesium is widely used since it is less expensive and, moreover, it introduces less impurity than does calcium[11]. In France, however, the reduction with calcium by bomb reaction is the preferred procedure[13].

After cooling, the uranium regulus is separated from the slag, and remelted in a vacuum induction-furnace[11]. It is vacuum cast into rods in graphite moulds. During the remelting some pick-up of impurities, such as carbon and nitrogen, may occur.

The reactions, briefly indicated here, are basic to uranium technology and will be discussed in more detail in later chapters. In recent years, the production of new and varied types of uranium fuel materials for nuclear reactors, designed for the economic generation of electrical power, have been developed. Among these are ceramic carbide and uranium nitride, UC and UN respectively, both of the natural and the enriched forms of uranium. Their properties are compared in Chapter 13 (p. 221).

Specification of purity

In a sense, purity is a relative term. The materials present which must be considered as impurities depend on the final purpose for which the uranium under preparation is required. From the point of view of a nuclear fuel, uranium and its compounds must not contain elements of high neutron absorption cross-section. These elements are, for instance, boron, cadmium, gadolinium, hafnium

TABLE 4

SPECIFICATION OF NUCLEAR GRADE URANIUM*

| Element | Eldorado | | General | ASTM[15] |
	maximum	typical	Electric	
Aluminium				75.0
Boron	0.2	0.1	0.2	0.25
Carbon	300	200	400	650
Calcium				50.0
Cadmium			0.2	0.25
Chlorine			5	
Chromium	65	10	20	50.0
Iron	150	60	50	175
Magnesium	25	8	5	25.0
Manganese	25	5	13	25.0
Nitrogen			50	100.0
Nickel	35	30	40	200.0
Silicon	65	35	50	120.0
Silver			1	

*All amounts are given in ppm parts of uranium.
For non-nuclear use, the restrictions by ASTM are:
| carbon | 750 ppm | iron | 200 ppm |
| chromium | 200 | nickel | 200 |

and samarium. The amount of these elements present should not exceed a few tenths of one part per million parts of uranium. From a mechanical standpoint the contents of other elements in the metal itself must be limited on account of their influence on the workability and methods of fabrication. The various requirements lead to the specifications set out for "nuclear grade" uranium of which typical examples are shown in Table 4.

It should be noted that impurity amounts are given on a uranium basis. Furthermore, the limits express the total permissible amounts of the elements, regardless of the chemical form in which they are present. The total impurities must not exceed 2000 ppm parts of uranium. It is now usual to express the impurities in terms of the boron equivalent, i.e. ppm boron equivalent to ppm element.

REFERENCES

1 J. D. WILSON, R. K. WEBSTER, G. W. C. MILNER, G. A. BARNETT AND A. A. SMALES, *Anal. Chim. Acta*, 23 (1960) 505.

2 *Gmelins Handbuch der Anorganischen Chemie*, 8. Auflage, *Uran und Isotope*, Syst. Nr. 55 (1936).

3 S. H. U. BOWIE, *Proceedings of the third international Conference on the Peaceful Uses of Atomic Energy, Geneva (1964)*, Vol. 12 (1965), p. 28.

4 P. F. KERR, *Proceedings of the first International Conference on the Peaceful Uses of Atomic Energy, Geneva (1955)*, Vol. 6 (1956), p. 5.

5 J. J. KATZ and E. RABINOWITCH, *The Chemistry of Uranium*, Part I, McGraw-Hill Book Company Inc., New York (1951).

6 S. J. W. GRIFFITH, *The Uranium Industry—Its history, technology and prospects*. Mineral Report 12, Department of Energy, Mines and Resources, Ottawa (1967).

7 *World Uranium and Thorium Resources*, ENEA, Paris (1965).

8 *Processing of Low-grade Uranium Ores, Proceedings of a Panel*, IAEA, Vienna (1967).

9 R. V. DAVIES, J. KENNEDY, R. W. McILROY, R. SPENCE AND K. M. HILL, *Nature*, 203 (1964) 1110.

10 E. T. PINKEY, W. LURIE AND P. C. N. VAN ZIJL, *Chemical Processing of Uranium Ores*, IAEA, Vienna (1962), Review series, No. 23.

11 W. D. WILKINSON, *Uranium Metallurgy*, Vol. I: *Uranium Process Metallurgy*, Interscience Publishers, New York, London (1962).

12 Ch. D. HARRINGTON AND A. E. RUEHLE, *Uranium Production Technology*, Van Nostrand Company Inc., New York (1959).

13 H. HUET, *Nuclear Power*, No. 48, April 1960.

14 J. H. GITTUS, *Uranium*, Butterworths, London (1963).

15 *The 1965 Book of ASTM Standards*, Part 7, ASTM (1965), p. 774.

CHAPTER 3

Uranium Metal

Introduction

Uranium is a soft, silvery-white metal. Its atomic number is 92 and its atomic weight is 238.03 ; thus it is the heaviest element found. Uranium falls in the actinide series of elements, a terrestrial series in which the 5f shell is in the course of being filled. The rather complicated electronic configuration of uranium is assumed to involve the valence electrons $5f^3 6d 7s^2$, and their excitation gives rise to the spectral features of the element[1]. However, the energy levels of the 5f, 6d and 7s shells are so close together that the magnetic and spectroscopic measurements which have been made on the element by different workers are not always in good accord. Moreover there is hybridization of the 5f and 6d electrons, as has been observed recently[2].

Natural uranium consists of a mixture of three isotopes with the atomic weights shown; the abundances are given in parenthesis: 238 (99.276 at.-%) 235 (0.718 at.-%) and 234 (0.0056 at.-%). The isotope ^{235}U, being fissile with thermal neutrons, is the present source of power in nuclear energy. The isotopes ^{238}U and ^{234}U are only fissile with fast neutrons and may therefore be applicable in fast reactors. Uranium has three allotropic modifications (α, β and γ). Their crystal lattices are of interest since they are important in determining the properties of the metal. Indeed, the phase transformations have had to be taken into consideration in the design of metallic fuel-elements. The physical properties of uranium, in particular the thermal conductivity, are of fundamental importance in its nuclear application. The properties of the metal are summarized in Table 5.

References p. 37

Uranium is a strongly electropositive element and its compounds are thus difficult to reduce to the metal. Correspondingly, uranium is very reactive; it combines more or less readily with all the elements.

With the metals it forms intermetallic compounds of great variety. The uranium alloys, which will be discussed in Chapter 4, are of considerable importance because they possess much better mechanical properties than those of uranium itself. For instance, hardness is markedly increased by alloying, and corrosion resistance and the irradiation stability are greatly improved by suitable additions.

TABLE 5

PHYSICAL PROPERTIES OF URANIUM

		Reference
Atomic weight	238.03	
Density (25 °C)		
X-ray	19.214	3
exp.	19.05 ± 0.02	6
Phase transformations		
$\alpha \rightarrow \beta$	667.7 °C	11
$\beta \rightarrow \gamma$	774.8 °C	
Melting point	1132.3 °C	11
Heat of fusion	2500 cal/g atom	15
	2900 cal/g atom	19
Heat of sublimation (0 °K)	129.0 kcal/mole	24
Specific heat	6.594 cal/°mole	15
Thermal conductivity (25 °C)	0.060 cal/cm. sec. deg	42

Crystallographic properties

Uranium has three allotropic modifications α, β and γ, between room temperature and its melting point.

α-*Uranium* has an orthorhombic structure with lattice parameters $a = 2.854$ Å, $b = 5.869$ Å and $c = 4.955$ Å[3]. The space group is Cmcm with four uranium atoms per unit cell at $0, y, 1/4$; $0, \bar{y}, 3/4$; $\frac{1}{2}, (\frac{1}{2}+y), 1/4$; $\frac{1}{2}, (\frac{1}{2}-y), 3/4$; The structure can be described in terms of corrugated sheets of atoms (Fig. 3); these lie parallel to the *b*-face in which the *y*-parameter is a measure of the degree of corrugation. The value of *y* is 0.1024 at room temperature[3]; its dependence on temperature has also been discussed[4]. The atomic bonding between atoms in the corrugated planes is greater than that between atoms in different planes.

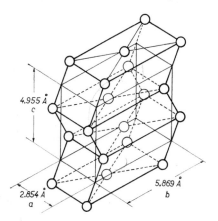

Fig. 3. Structure of α-uranium.

Some similarity between the structure of α-uranium, in which a high degree of covalent bonding exists, and covalent compounds such as PCl_5 has been noted[5]. The structure shows a degree of complication but has the unusual feature that there is contraction along the *b*-axis upon heating (Fig. 4). This anomaly, together with the fact that the coefficients of thermal expansion in the other two axial directions are high, causes difficulties in the thermal cycling of the metal (see p. 36).

The X-ray density is 19.214; the density measured from high-purity, directionally solidified metal is 19.05 ± 0.02[6]. A method of growing perfect crystals of α-uranium has been described recently[7].

β-Uranium has a tetragonal structure with lattice parameters (at 700°C) of $a = 10.754$ Å and $c = 5.623$ Å; the X-ray density is 18.13[8]. The complicated structure, which is not yet fully elucidated[9], has a large number of atoms in the unit cell. The β-phase is important in the production of uranium metal since heat treatment within the β-phase temperature range is employed to destroy the preferred orientation created during its fabrication. The α–β

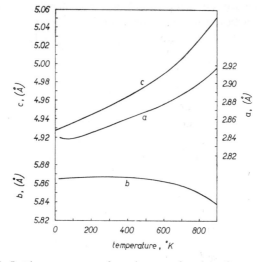

Fig. 4. Lattice parameters of uranium as a function of temperature

transition in pure uranium has been investigated with the aid of high-temperature X-ray analysis[10]. This showed that the β-crystallites become much larger than the parent α-crystallites, even after having passed through the transition for the first time.

γ-*Uranium* has a simple body-centered cubic structure with $a = 3.53$ Å at $800\,°C$[8].

The phase diagram

The phase transformations in uranium metal have been the subject of many investigations, since their occurrence seriously limits the use of the metal as a fuel in nuclear reactors. Several methods of studying the transformations have been used, including the measurement of conductivity, dilatometry and high-temperature X-ray analysis. These measurements show that the α-modification is transformed into the β-form at $667.7\,°C$ and the β-modification, in turn, into the γ-form at $774.8\,°C$[11]. The last phase has a melting point of $1132.3\,°C$.

Just as they modify other physical properties, impurities in the metal strongly affect the uranium transition temperatures.

In several instances, for example by alloying with zirconium, the β–γ transition is so much depressed that, on cooling, the γ-phase is transformed directly into α-uranium. This has found an application in nuclear technology where it is used to increase the dimensional stability of the thermal-cycled uranium (see Chapter 4). The effect of high pressures and temperatures on the phase relationships in pure uranium has been reported[12,13]. The triple point, at which α, β and γ are in equilibrium, is at 798 °C under a pressure of 29.8 kbar[14]. At pressures above 35 kbar, the γ-modification is transformed directly into the α-form. The phase diagram is shown in figure 5.

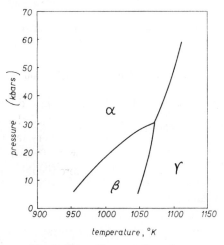

Fig. 5. Phase diagram of uranium.

Thermodynamic properties

The thermodynamic properties of pure uranium metal have been assessed by Rand and Kubaschewski[15] and their data are used here, unless it is indicated otherwise. The heat capacity of α-uranium, as measured independently by several authors, whose results are in excellent agreement[16,17,18], may be represented by the equation:

$$C_p = 2.61 + 8.95 \cdot 10^{-3} \cdot T + 1.17 \cdot 10^5 \cdot T^{-2} \text{ cal/°mole}$$
$$(298\text{–}941 \text{ °K})$$

The heat capacities of β-uranium and γ-uranium appear to be independent of temperature and are 10.0 and 9.1 cal/°mole respectively. The entropy of α-uranium follows from low-temperature heat-capacity measurements[19,20]; at 298 °K $S° = 12.0$ cal/°mole. For the heats of transition it has been found[15,18]:

$$\alpha \rightarrow \beta + 750 \text{ cal/g. atom}$$
$$\beta \rightarrow \gamma + 1150 \text{ cal/g. atom}$$

The heat of fusion has been estimated[15] from the freezing-point depression produced by dissolving various metals in uranium. The result, $-\Delta H = 2500 \pm 500$ cal/atom, is in good agreement with the value, 2900 cal/g. atom, found in a recent determination[18]. The vapour pressure of liquid uranium has been measured by Rauh and Thorn[21] and by Alcock and Grieveson[22] using the Knudsen effusion technique. However, Drowart et al.[23], using a Knudsen cell in combination with a mass spectrometer, have shown that these pressures are too high and correspond to the vapour pressures over the equilibrium $U(l) + UO_{2-x}(s)$, as will be discussed in more detail later (p. 90).

Obviously, knowledge of the contribution of the molecules $UO(g)$ and $UO_2(g)$ to the total pressure is important since it is virtually impossible to avoid oxidation of uranium when samples are prepared. Thus, for the uranium partial pressure, Drowart et al.[24] found:

$$\log p(\text{atm}) = \frac{-26,210}{T} + 5.920 \qquad (1720\text{–}2340 \text{ °K})$$

From these results a heat of sublimation of uranium, $-\Delta H_0 = 129.0$ kcal/g. atom, has been derived.

Preparation of uranium

Uranium metal was first prepared by Péligot, who reduced uranium tetrachloride with potassium. From that time until about 1940 uranium was made on a small scale only, several methods being used[25]. These included the reduction of uranium dioxide with carbon or strongly electropositive metals such as aluminium or calcium and the electrolytic reduction of uranium halides in fused-salt baths. Such methods now have only historic interest and are excellently reviewed by Katz and Rabinowitch[25]. Since 1940 production of the metal on a larger scale has become important as a consequence of the development of nuclear energy and other methods have been employed. The reduction of uranium oxide with carbon, which might seem attractive, is not feasible unless the temperature is very high, somewhat over 2000 °C. Only then is the equilibrium pressure of the reaction:

$$UO_2 + C \rightleftarrows U(l) + CO$$

high enough to remove the gas sufficiently quickly to allow the reaction to proceed to the right at a practically useful rate. Moreover the uranium produced dissolves carbon with the formation of carbides. Reduction of the dioxide with hydrogen is, as might be expected, even more difficult to achieve.

The only method of technological importance is one involving the reduction of a uranium halide. Uranium tetrafluoride, UF_4, is commonly used, although the chloride, UCl_4, also gives a thermodynamically feasible route. However, since the latter compound not only deliquesces, but oxidizes and hydrolyses easily in air, it is technologically not an attractive starting material because of handling difficulties.

In practice, the tetrafluoride has been the exclusive choice and the reducing agent has been either magnesium or calcium:

$$UF_4 + 2\,Mg \rightarrow U + 2\,MgF_2 + \ \ 82\,kcal$$
$$UF_4 + 2\,Ca \rightarrow U + 2\,CaF_2 + 134\,kcal$$

Magnesium is generally favoured because of its comparative

cheapness and the ease with which it can be handled. However, because of the lower heat of reaction with this metal, the mixture of UF_4 and magnesium has to be heated to a temperature of about 600–700 °C to initiate the reaction. The heat liberated by the reduction raises the temperature to a level at which the uranium melts and collects in the lower parts of the crucible, under a slag of MgF_2. After cooling the billet of uranium may be removed.

Uranium of exceptionally high purity can be made by the thermal decomposition of uranium tetraiodide, UI_4, on a tungsten filament. The preparation, essentially a laboratory method, is only possible on a small scale and is favoured by a low partial pressure of iodine[26]. This avoids the dissociation of UI_4 in the vapour phase into the non-volatile UI_3. The kinetics of the process have been examined by Hashino and Kawai[27]. From this it appeared that the uranium transport was only *via* the tetraiodide phase, while the formation rate of UI_3 was so slow that it did not substantially participate in the process.

Fabrication of uranium

The fabrication of uranium[28,29], as outlined above, is carried out in a closed reaction vessel or bomb which is lined with MgF_2 and is electrically heated slowly over a 12–16 hour period, to a temperature controlled at about 650 °C.

Uranium tetrafluoride ("green salt") and magnesium are delivered in small, weighed drums, ready for use. Magnesium is used in an excess to the extent of about 6%. The charges of UF_4 and magnesium are blended and introduced into the bomb by means of a filling machine. The bomb is then heated until the charge ignites, after which the reduction to uranium metal proceeds exothermically. The ignition temperature is about 600 °C.

The composition of the green salt, essentially UF_4, has a significant effect on the firing time, the yields and the quality of the metal produced. Normally the green salt contains small amounts of uranyl fluoride and dioxide, the presence of both of which are unfavourable to good yields. Of the two, oxide is the less desirable since it produces MgO which increases the viscosity of the slag. This makes separation in the bomb of metal from slag more difficult. The best results are obtained with UF_4 of more than 98% purity; then the yields are 97%.

Following the reaction, the bomb is cooled, first in air to 500 °C, then by immersion in water up to a point a little below the shell flange. After the regulus has been manually separated from the slag, the crude metal is further cleaned from adhering slag with an air-operated hammer. Surface impurities are removed by pickling the regulus in 40% nitric acid at 80 °C. The regulus after having been washed and dried, is melted in a vacuum furnace at 1450 °C where it is outgassed. The molten uranium is cast in a graphite mould, situated directly under the crucible. After the melt is poured into the mould, it is cooled in a stream of inert gas (argon or helium) until a temperature of about 300 °C is reached. At this temperature oxidation of the metal does not take place and the furnace is brought to atmospheric pressure with nitrogen and opened. The loaded mould is stripped from the ingot and air oxidation is minimized by immersion in water. The cooled ingot is then weighed and its top, which forms the discard and is higher in impurities, is removed by sawing.

There is an advantage in avoiding the vacuum remelting step and obtaining directly the ingot, called "dingot". This process furnishes a product of a higher purity at lower costs than that obtained by reduction-casting. However, the hydrogen level may be higher than in ingots, owing to impurities of water present in the starting materials. When the hydrogen level exceeds a maximum value of 2 ppm, it causes serious embrittlement of uranium metal. The dingot is readily separated from the slag by hammering. It is then cleaned and weighed to determine the crude yield. All surfaces are machined to remove external layers of impurities, particularly from the top where there is a concentration of impurities and slag. The metal removed in this way is collected and remelted. After the machined dingot has passed the control test and has been weighed, it is ready for further processing into the required shapes by either forging or extrusion. For this purpose, the dingot is preheated for three hours in a molten salt bath, which is held below the temperature of the α–β transformation; it is then forged in a forge press into shapes suitable for secondary rolling or extrusion to final size

Enriched uranium

Industrially the isotopic enrichment of uranium is achieved by gas diffusion through porous membranes. Uranium hexafluoride, UF_6, produced by the fluorination of uranium tetrafluoride at about 300 °C, is the gas used. The isotopes ^{238}U and ^{235}U diffuse through the membranes (pores $\sim$ 100 Å) at a slightly different rate. To attain a high enrichment many diffusion stages are needed. The enriched uranium hexafluoride finally obtained, is reduced to the tetrafluoride by means of hydrogen. The enriched metal may then be prepared by the reduction of the enriched uranium tetrafluoride to metal in the way already described for the ordinary material.

The centrifugal separation of the fluorides, which is likely to displace diffusion in time, is done in principle in a long, narrow vertical cylinder rotating about its axis with high angular velocity. The mixture of the gaseous fluorides, contained in the cylinder, will tend to separate, with the component of higher molecular weight concentrating towards the cylinder walls and the component of lower molecular weight towards the axis. By moving the heavy stream near the outside in a direction axially opposite to that of the light stream, it is possible to establish a longitudinal composition gradient which, if the machine is long enough, or conversily has sufficient stages, will provide any desired difference in composition of the isotope mixture.

Chemical properties

Uranium readily reacts with most non-metals at temperatures, depending on the degree of subdivision. For instance, with massive metal the temperatures required are for nitrogen > 400 °C, carbon

> 1250 °C, and sulphur > 250 °C. But with finely divided uranium the corresponding temperatures are much lower. These and other reactions will be discussed at length in the appropriate chapters.

From a practical point of view, the reaction of uranium with oxygen or oxidizing gases is naturally of great interest. Finely divided uranium is higly pyrophoric; it ignites spontaneously in air at room temperature and sometimes even in water. It is also to be noted that powders of other metals which are mixed with uranium powder in the making of alloys may also be pyrophoric and potentially explosive in air.

When the massive metal is exposed to air, its surface loses its silvery lustre and becomes covered with an oxide layer, which effectively protects it from further oxidation at room temperature and slightly above. The oxide layer varies from yellow to black, according to its thickness. The kinetics of oxidation of the massive metal have been examined by Cubicciotti[30].

When heated in air, uranium ignites at temperatures depending on the size of the particles, but the massive metal burns slowly but steadily at 700 °C to form the oxide. Uranium turnings burn brilliantly in oxygen, when ignited at 700–1000 °C, emitting a white light. The metal is also oxidized by carbon dioxide in the temperature range of 400–650 °C. The rate is controlled by the diffusion of O^{2}-ions through a thin layer of uncracked UO_2 which is maintained on the surface of the metal[31].

Uranium is readily oxidized by water even at room temperature, but the reaction is inhibited by oxygen. The kinetics of the oxidation with water vapour between 35–100 °C has been examined by Baker *et al.*[32]. The reaction:

$$U + 2 H_2O_{(g)} \rightarrow UO_2 + 2 H_2$$

occurs at linear rate, depending on the surface area and on the water vapour pressure, but not on the hydrogen pressure. In the course of the reaction uranium hydride, UH_3, is formed. Of this a part is oxidized by water:

$$UH_3 + 2 H_2O \rightarrow UO_2 + 3\tfrac{1}{2} H_2$$

the rest is protected by the uranium oxide formed[32]. The oxide layer ($UO_{2.06}$) produced by the action of water on the hydride does not form a protective layer and a mechanism has been proposed in which the rate determining step is the diffusion of water to the internal oxide surface. Oxygen dissolved in the water was found to reduce the corrosion rate in the uranium/water system; the inhibition depends on the oxygen partial pressure[33]. The corrosion resistance of uranium to water at elevated temperatures is much increased by alloying the uranium with niobium or molybdenum.

Uranium in the mass reacts only slowly with dilute mineral acids, such as sulphuric acid or phosphoric acid; it is remarkably resistant to dilute sulphuric acid (6 N), even when the acid is boiled. However, uranium may be dissolved in H_2SO_4 with the aid of hydrogen peroxide, and it dissolves slowly in hot, concentrated sulphuric acid. Other acids, such as H_3PO_4, react only slowly with uranium in the cold, forming uranous salts.

The dissolution behaviour of uranium in acids is rather unusual since, unlike other metals, the rate decreases with increasing acidity. It has been shown[34] that hydrogen ions inhibit dissolution. Dissolution is most effectively achieved in hot, dilute nitric or sulphuric acid, especially when soluble oxidizing agents, such as perchloric acid or hydrogen peroxide, are present. The reaction of uranium with hydrochloric acid is of particular interest. Finely divided metal dissolves violently in dilute acid, but the rate of dissolution of the massive metal is slower and depends on the acid concentration. With concentrated HCl it is very rapid; dilute HCl, say 6 N, still reacts quite rapidly, but when the dilution reaches 1 N, the rate is slow[35]. The reaction is clearly not a simple one in that the dissolution is accompanied by the formation of a black precipitate (UO_2?). Moreover, the uranium in the solution produced appears in both the $+3$ and the $+4$ states; the ratio of these ions is variable, depending on the acid concentration, the temperature and time of dissolution. The reaction would appear to be:

$$U + 3\,HCl \rightarrow UCl_3 + 1\tfrac{1}{2}\,H_2$$

followed by:

$$UCl_3 + HCl \rightarrow UCl_4 + \tfrac{1}{2}\,H_2$$

From measurements of the hydrogen evolved, it has been found that the average valence state is somewhere between 3.2 and 3.4 when the dissolution is in 6 N HCl. In 12 N HCl, however, almost all the uranium going into solution is in the $+4$ state[35].

With oxidizing acids uranium dissolves rapidly. Especially with nitric acid the reaction is of considerable interest, since this acid is used to bring spent uranium fuel into solution for the purpose of reprocessing. The stoichiometry of the reaction depends on the acid concentration; in 13 N HNO$_3$ it is:

$$U + 5.5\ HNO_3 \rightarrow UO_2(NO_3)_2 + 2.25\ NO_2 + 1.25\ NO + 2.75\ H_2O$$

whereas in 8 N HNO$_3$ it is:

$$U + 4\ HNO_3 \rightarrow UO_2(NO_3)_2 + 2\ NO + 2\ H_2O$$

Uranium is inert to alkali solutions; only the addition of oxidizing agents, such as that of sodium peroxide to sodium hydroxide solutions, leads to dissolution.

Interesting is the displacement of several metals from solutions of their salts by uranium. For instance, uranium dissolves in a copper(II) acetate solution. Silver acetate reacts only slowly, since the uranium surface becomes coated with a protective layer of silver, but with a solution of copper(II) sulphate there is no reaction[35].

Physical properties

The physical properties of the pure, unalloyed metal demand attention since they allow a better understanding to be obtained of the metallurgical behaviour of the metal. In α-uranium the highly anisotropic structure is reflected in its physical and mechanical properties. These are not only affected by purity but also by the previous metallurgical history of the metal.

Electrical properties

Uranium is a rather poor conductor of electricity, comparable in this respect with iron. Resistivity measurements on α-uranium, between 97–293 °K and in the three crystallographic directions have been reported by several authors[36,37,38], with results which are in a fair agreement (Table 6). It was found that the direction showing the largest thermal expansion also shows the largest electrical resistivity.

TABLE 6

RESISTIVITY OF α-URANIUM AT ROOM TEMPERATURE (25°C)

	Berlincourt[36]	Pascal[37]	Raetsky[38]
	$\sigma(\mu$ ohm cm)		
[100]	39.4	41	36.4
[010]	25.5	25.1	25.1
[001]	26.2	32	32.1

The Hall coefficient has a positive value, $+3.9 \cdot 10^{-5}$ cm^3/Coul, corresponding to a theoretical concentration of three conduction electrons per atom[36]. Uranium is reported to become superconducting below 0.68 °K[39]. Values for the absolute thermoelectric EMF have been reported by Tyler et al.[40].

Uranium is weakly paramagnetic. The temperature dependence of the magnetic susceptibility has been measured by Bates and Hughes[41].

The thermal conductivity has been examined recently by Howl[42]. From a large number of measurements a smooth increase in the thermal conductivity of uranium with temperature has been found; the relationship is according to the equation:

$$K = 0.0585 + 6.55 \cdot 10^{-5} \cdot t$$

(K in cal/cm. sec. deg and t in °C). The measurements are in fair agreement (better than 4%) with the older measurements of Smith[43]. The influence of small amounts of impurities, such as the aluminium or iron found in "adjusted" uranium, appears to be negligible[42].

The thermal expansion has been measured by X-ray lattice parameter measurements on polycrystalline samples[8,44]. Although the results are in good agreement, some doubts have been cast on their reliability[45] because of the restraints imposed by the individual grains upon the thermal expansion of their neighbours. The most reliable measurements have been made on carefully grown single crystals, free from sub-boundaries[45]. The linear dilatation equations, based on these measurements are:

[100] $l_t = l_0(1 + 23.53 \cdot 10^{-6} \cdot t + 13.74 \cdot 10^{-9} \cdot t^2 + 9.94 \cdot 10^{-12} \cdot t^3)$

[010] $l_t = l_0(1 + 1.16 \cdot 10^{-6} \cdot t - 9.43 \cdot 10^{-9} \cdot t^2 - 11.79 \cdot 10^{-12} \cdot t^3)$

[001] $l_t = l_0(1 + 19.38 \cdot 10^{-6} \cdot t + 21.58 \cdot 10^{-9} \cdot t^2 + 3.32 \cdot 10^{-12} \cdot t^3)$

The volume dilatation can be represented by the equation:

$$V_t = V_0(1 + 43.98 \cdot 10^{-6} \cdot t + 26.88 \cdot 10^{-9} \cdot t^2 + 1.00 \cdot 10^{-12} \cdot t^3)$$

Mechanical properties

Uranium is a soft metal but its pronounced tendency to work-hardening makes it difficult to machine[46]. Heat treatment in the alpha temperature range is given to remove the effects of work-hardening. The mechanical properties of the highly anisotropic metal are markedly affected by orientation. A uranium casting made by pouring metal into a graphite mould will have relatively large grains and variable mechanical properties. This is also true of the metal after most mechanical operations[47].

Accordingly, the α–β transformation in uranium has been studied[48] in relation to a way of attaining complete texture-randomization on the subsequent beta heat treatment. It has been found that specimens quenched in water were less anisotropic after a given time at a certain temperature in the beta-temperature range than was air-cooled material. The beta heat-treatment, in combination with suitable alloying (see Chapter 4), helps to randomize the grain structure in uranium.

The mechanical properties of uranium are also influenced by its purity. For instance, hydrogen has a harmful effect on the ductility of the metal: even small amounts, in the 0.3–5 ppm range, make it brittle.

On thermal cycling, that is repeated heating and cooling through the α–β transformation temperature, uranium undergoes growth in volume as well as cracking and severe surface roughening. These dimensional changes are attributed to the anisotropic coefficient of expansion of α-uranium which causes stress to be developed across the boundaries of the adjoining grains. The deformation depends on a number of factors, including the cooling rate and the amount of alloying material. Thus it has been observed[49] that as the rate of cooling is increased, the growth increases markedly. Alloying additions, such as Nb, Cr, Mo or Zr, give metal showing a considerable improvement over unalloyed uranium (see Chapter 4).

Irradiation behaviour of uranium

When uranium is irradiated with neutrons, it tends to distort and to swell. Much effort has been devoted to evolving carefully controlled methods for the casting and heat treatment of the metal to ensure the metallurgical properties needed to minimize these effects. Several mechanisms play a part and the effects which they produce are dependent on the conditions of irradiation. Thus at temperatures below 400 °C anisotropic growth occurs, whereas above 500 °C this does not occur[50].

Irradiation of unalloyed uranium in the high alpha-temperature range (400–670 °C) is accompanied by significant and variable volume increase. This swelling is a major concern in reactor design and operation since in some cases it may contribute to fuel failure. A further cause of trouble is that solid fission atoms occupy more volume than did the fissioned uranium atoms. The swelling thus occasioned is independent of the irradiation temperature and is 3 % per atom percentage burn-up of uranium. When the irradiation temperature has been over 300 °C, the fission product gases xenon and krypton agglomerate and form small bubbles visible in replicas taken from irradiated metal as voids of 0.01 to 1.0 μ in diameter. A volume increase of several percent may be caused by the gas bubbles, the increase becoming greater as irradiation temperature and burn-up is increased.

A further mechanism contributing to swelling is that described as mechanical cavitation[51]; it is the principal cause of swelling at temperatures between 370 and 500 °C. An internally generated stress can be the driving force leading to cavitation. A source of this stress is the anisotropic expansion of the individual grains during thermal cycling. The cavitation is primarily at grain and twin boundaries.

The effects of swelling and creep described above may be minimized by the use of alloying elements and by employing high strength cladding over the metal fuel element.

REFERENCES

1 J. J. KATZ AND E. RABINOWITZ, *The Chemistry of Uranium*, McGraw-Hill Book Company, Inc. New York (1951), p. 17.
2 B. ROSENFELD AND M. SZUSZKIEWICZ, *Nukleonika*, 11 (1966) 693.
3 E. F. STURCKEN, *Acta Cryst.*, 13 (1960) 852.
4 H. MUELLER, R. L. HITTERMAN AND H. W. KNOTT, *Acta Cryst.*, 15 (1962) 421.
5 L. PAULING, *The Nature of the Chemical Bond*, 2nd Ed., Cornell University Press, Ithaca, N.Y. (1948), p. 413.
6 B. BLUMENTHAL, *ANL-Report* 5019 (1953).
7 F. JEAN-LOUIS, J. S. DANIEL AND P. LACOMBE, *Compt. Rend.* [C], 264 (1967) 1818.
8 H. H. KLEPPER AND P. CHIOTTI, *ISC-Report* 893 (1957).
9 H. STEEPLE AND T. ASWORD, *Acta Cryst.*, 21 (1966) 995.
10 J. EISENBLÄTTER, G. HAASE AND F. GRANZER, *J. Nucl. Mat.*, 21 (1967) 53.
11 B. BLUMENTHAL, *J. Nucl. Mat.*, 2 (1960) 23.
12 W. KLEMENT, Jr., A. JAYARAMAN AND G. G. KENNEDY, *Phys. Rev.*, 129 (1963) 1971.
13 L. T. LLOYD, R. G. LIPTAI AND R. J. FRIDDLE, *J. Nucl. Mat.*, 19 (1966) 173.
14 N. ASAMI, M. YAMADA AND S. TAKAHASHI, *Nippon Kinzoku Gakkaishi*, 31 (1967) 389.
15 M.H. RAND AND O. KUBASCHEWSKI, *The Thermochemical Properties of Uranium Compounds*, Oliver and Boyd, Edinburgh and London (1963), p. 8.
16 G. E. MOORE AND K. K. KELLEY, *J. Am. Chem. Soc.*, 69 (1947) 2105.
17 D. C. GINNINGS AND R. J. CORRUCCINI, *J. Res. Nat. Bur. Stand.*, 39 (1947) 309.
18 H. SAVAGE AND R. D. SEIBEL, *ANL-Report* 6702 (1963).
19 W. M. JONES, J. GORDON AND E. A. LONG, *J. Chem. Phys.*, 20 (1952) 695.
20 H. E. FLOTOW AND H. R. LOHR, *J. Phys. Chem.*, 64 (1960) 904.
21 R. J. ACKERMANN, E. G. RAUH AND R. J. THORN, *J. Chem. Phys.*, 37 (1962) 2693.

22   C. B. ALCOCK AND P. GRIEVESON, *J. Inst. Met.*, 90 (1962) 304.

23   J. DROWART, A. PATTORET AND S. SMOES, *J. Chem. Phys.*, 42 (1965) 2629.

24   A. PATTORET, J. DROWART AND S. SMOES, *Trans. Faraday Soc.*, 65 (1969) 98.

25   J. J. KATZ AND E. RABINOWITZ, see ref. 1, p. 122.

26   C. H. PRESCOTT, F. L. REYNOLDS and J. A. HOLMES, *MDDC-Report* 437 (1946).

27   T. HASHINO AND T. KAWAI, *Trans. Faraday Soc.*, 63 (1967) 3088.

28   Ch. D. HARRINGTON AND A. E. RUEHLE, *Uranium Production Technology*, Van Nostrand Company, Inc. (1959).

29   W. D. WILKINSON, *Uranium Metallurgy*, Interscience Publishers, Vol. I, *Uranium Process Metallurgy* (1962).

30   D. CUBICCIOTTI, *J. Am. Chem. Soc.*, 74 (1952) 1079.

31   J. J. STOBBS AND J. WHITTLE, *J. Nucl. Mat.*, 19 (1966) 160.

32   M. McD. BAKER, L. N. LESS AND S. ORMAN, *Trans. Faraday Soc.*, 62 (1966) 2513.

33   M. McD. BAKER, L. N. LESS AND S. ORMAN, *Trans. Faraday Soc.*, 62 (1966) 2525.

34   L. E. KINDLIMANN AND N. D. GREENE, *Corrosion Sci.*, (1967) 29.

35   J. C. WARF, in: *The Chemistry of Uranium*, Collected papers, edited by J. J. KATZ AND E. RABINOWITZ. TID-Report 5290 (1958), p. 29.

36   T. G. BERLINCOURT, *NAA-SR-Report* 1885 (1956).

37   J. PASCAL, J. MORIN AND P. LACOMBE, *J. Nucl. Mat.*, 13 (1964) 28.

38   V. M. RAETSKY, *J. Nucl. Mat.*, 21 (1967) 105.

39   R. A. HEIN, W. E. HENRY AND N. M. WOLCOTT, *Phys. Rev.*, 107 (1957) 1517.

40   W. W. TYLER, A. C. WILSON, Jr. AND G. J. WOLGA, *KAPL-Report* 802 (1952).

41   L. F. BATES AND D. HUGHES, *Proc. Phys. Soc.*, B 67 (1954) 28.

42   D. A. HOWL, *J. Nucl. Mat.*, 19 (1966) 9.

43   K. F. SMITH, *ANL-Report* 5700 (1957).

44   J. R. BRIDGE, C. M. SCHWARTZ AND D. A. VAUGHAN, *Trans. AIME*, 206 (1956) 1282.

45   L. T. LLOYD, *ANL-Report* 5972 (1959).

46   W. D. WILKINSON, ref. 29, p. 440.

47   S. F. PUGH, *J. Inst. Met.*, 86 (1958) 497.

48   K. M. PICKWICK AND W. J. KITCHINGMAN, *J. Nucl. Mat.*, 19 (1966) 109.

49   R. SOMASUNDARAM AND V. PREMANAND, *J. Nucl. Mat.*, 19 (1966) 283.

50   S. N. BUCKLEY, *AERE-R-Report* 3674 (1961).

51   C. L. ANGERMAN AND G. R. CASKEY, Jr., *J. Nucl. Mat.*, 13 (1964) 182.

CHAPTER 4

Interaction of Uranium with Metals

Introduction

Metallic uranium reacts readily with many metals and nonmetals. The uranium alloys are of considerable interest in nuclear technology since they may have several advantages over the unalloyed metal. Uranium itself has properties which are, in many respects, disadvantageous; it is active chemically, it has an anisotropic structure and its mechanical properties are poor. The purpose of alloying is to improve the behaviour of the parent metal. For instance, the swelling and creep of unalloyed uranium under neutron irradiation at temperatures above 400 °C may be minimized by alloying.

The properties of uranium alloys of particular interest in nuclear technology are thermal conductivity, crystallographic and metallographic structures, heat-treatment characteristics and irradiation stability. Detailed information on these properties is given in specialized monographs, for instance, in *Uranium Metallurgy* (Vol. II, 1962) by Wilkinson[1] and in *Uranium* (1963) by Gittus[2]. Therefore, only relevant information on binary systems will be given here, followed by a brief discussion of the application of alloys belonging to these systems in nuclear technology.

Uranium and metals of Group I

The systems formed by uranium with the *alkali metals* have been insufficiently investigated. On the contrary, the systems with members of the subgroup, copper, silver and gold, have been well examined. *Copper* forms the compound UCu_5 which has a face-centered cubic lattice[3] with $a = 7.033$ Å, and melts incongruently at about 1065 °C (Fig. 6). *Silver* does not form intermetallic compounds with uranium[4]. In the liquid state there is a considerable region of immiscibility, but the mutual solubility increases in range with increasing temperature. The solubility of *gold* in the γ-uranium phase is 6 at. %. The element forms the incongruently melting compounds UAu_2 (m.p. 1216 °C) and UAu_3[5].

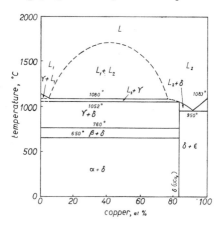

Fig. 6. Phase diagram of the copper–uranium system.

Thermodynamic properties of liquid alloys in the uranium–gold system over the composition range 17–90 at.% have been reported[6].

Uranium and metals of Group II

Beryllium gives with uranium the compound UBe_{13}. This has a face-centered cubic structure with $a = 10.256$ Å. Its heat of formation has been given as $-\varDelta H_{298} = 39.3$ kcal/mole[7]. There is a small miscibility gap in the phase diagram of the system between 97.5 and 99 at. % beryllium. The alloy has a good resistance to oxidation and corrosion at high temperatures.

Magnesium does not form an intermediate compound with uranium and its solubility in that metal is very low: 0.002 wt. % Mg at 650 °C and 0.17 wt. % at 1132 °C[8]. The transition temperatures of uranium and its melting point are not affected. There is a very large miscibility gap in the system.

Calcium does not react with uranium up to 800 °C; even liquid calcium does not react significantly. Compounds have not been reported.

Zinc, like all the metals of the subgroup, forms compounds with uranium. Zinc gives a hexagonal compound, U_2Zn_{17}[9], and two, or possibly three, closely related compounds in the composition range UZn_{12} to U_2Zn_{17}[10]. The small difference in the thermodynamic stability of these phases is responsible for difficulties which have been encountered in establishing the phase diagram of this system.

The *cadmium*-uranium system is very interesting in that a retrograde solubility of uranium in liquid cadmium occurs between 473 °C and 630 °C[11]. The thermodynamics of solution as well as the free energy of formation of the only compound in the system, UCd_{11}, have been ascertained by means of EMF-measurements[12].

Mercury forms three compounds with uranium: UHg_2 being hexagonal with $a = 4.986$ Å and $c = 3.225$ Å; UHg_3, hexagonal with $a = 3.327$ Å and $c = 4.888$ Å, and UHg_4, body-centered cubic with $a = 3.63$ Å[13]. These compounds are all pyrophoric. Their thermodynamic properties have been dealt with by Forsberg[14], who measured the partial pressure of mercury over alloys in the three-phase regions. However, the interpretation of the measurements seems to be in doubt[10].

Uranium and metals of Group III

In the *boron*-uranium system the phases UB_2, UB_4 and UB_{12} have been found. The crystal structures of these borides have been studied by Bertaut and Blum[15] and by Zalkin and Templeton[16, 17]. The boride UB_2 is a hexagonal layer-type structure with $a = 3.13$ Å and $c = 3.99$ Å; the space group is C6/mmm. The sheets of uranium atoms, normal to the *c*-axis, are held together by the boron atoms, forming U–B–U bonds in addition to the B–B bonds in the network. The thermal expansion of UB_2 has been investigated with a high-temperature X-ray camera[18]. The mean value between room temperature and 205°C is $9 \cdot 10^{-6}/°C$ (*a*-axis) and $8 \cdot 10^{-6}/°C$ (*c*-axis).

The tetraboride UB_4 has a tetragonal unit cell with lattice parameters $a = 7.080$ Å and $c = 3.978$ Å; its space group is P4/mbm[19]. The dodecaboride UB_{12} has been prepared for the first time by Andrieux and Blum[20]. It has a face-centered cubic unit cell with $a = 7.477$ Å[21]; the space group is Fm3m. It is reported to be diamagnetic[19].

The uranium borides can be prepared by heating calculated amounts of boron and finely divided uranium (obtained by the decomposition of UH_3) in the form of cold pressed compacts in molybdenum crucibles under an argon atmosphere. The reaction between uranium and boron begins between 1000° and 1250°C and liberates much heat. The temperature is then raised to between 1400° and 2000°C, depending on the sample. The resulting product is a compact that is practically not sintered and can easily be powdered. During the preparation some oxygen may be introduced which is made evident by the presence of UO_2 in the product. When borides of high density have to be prepared, hot pressing in graphite dies is necessary, but UB_2 is reported to decompose partially under this treatment.

UB_4 was first obtained in a crystalline form by electrolysis from a molten salt bath. When a mixture of magnesium borate and magnesium fluoride, in which U_3O_8 is dissolved[22], is used, pure UB_4 crystals are formed at 1100°C during the passage of a current of 12 V and 23 A. The tetraboride can also be prepared by the reaction between boron and uranium tetrafluoride in an inert atmosphere, which takes place between 1000 and 1700°C, according to:

$$3 \text{ UF}_4 + 16 \text{ B} \rightarrow 3 \text{ UB}_4 + 4 \text{ BF}_3$$

The formation of UB_4 proceeds above 1250°C, but only above 1600°C will UB_4 be formed without UB_2. An interesting method for the preparation of UB_4 is the reaction between boron and uranium dioxide at 2000°C:

$$\text{UO}_2 + 6 \text{ B} \rightarrow \text{UB}_4 + (\text{BO})_2 \text{ (g)}$$

When this is carried out in graphite the product contains $\sim$ 10 wt. % carbon. In an electron bombardment melting furnace, however, a pure product can be obtained[23].

The phase diagram of the boron–uranium system (Fig. 7) has been given by Howlett[21]. UB_2 is likely to exist over a small range of

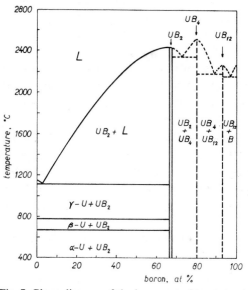

Fig. 7. Phase diagram of the boron–uranium system.

composition; it forms a simple eutectic with liquid uranium at 1108 °C and at 2–3 at. % boron. The melting point of UB_2 is 2240 °C; the borides UB_4 and UB_{12} melt at 2495° and 2235 °C respectively. The free energies of formation of the borides UB_2, UB_4 and UB_{12} have been determined by Alcock and Grieveson from vapour pressure measurements using a Knudsen effusion cell[24]. They obtained:

$$U(l) + 2B(s) \rightarrow UB_2(s) \qquad \Delta G_T^0 = -39,300 + 3.0\ T$$

$$U(l) + 4B(s) \rightarrow UB_4(s) \qquad \Delta G_T^0 = -60,400 + 4.4\ T$$

$$U(l) + 12B(s) \rightarrow UB_{12}(s) \qquad \Delta G_T^0 = -106,000 + 10.5\ T$$

The properties of the uranium borides have not been extensively studied. UB_{12} is not stable at high temperatures since it has a high boron partial pressure[24] and hence decomposes quickly. The stability of the diboride, UB_2, with respect to UB_4 perhaps decreases with increasing temperature above 1500 °C.

The oxidation of UB_2 has been examined by Albrecht and Koehl[25], who found a linear law up to 400 °C. Above that temperature it reacts anisothermally which results in complete combustion in a short time. There is no reaction with nitrogen below 1000 °C. UB_4, which is of interest as a burnable poison in nuclear fuel, is compatible with stoichiometric UO_2 up to the melting point of UB_4. However, a small excess of oxygen in UO_2 reacts readily with UB_4 to give gaseous $(BO)_2$[23].

The *aluminium*–uranium system is of particular interest, since alloys of high aluminium content, UAl_4, have found wide application as reactor fuels, particularly in water-cooled materials-testing reactors. The phase diagram of the system is given in Fig. 8

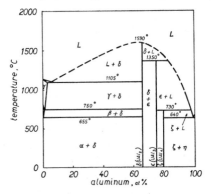

Fig. 8. Phase diagram of the aluminium–uranium system.

which shows that aluminium forms three compounds with uranium, UAl_2, UAl_3 and UAl_4[26]. UAl_2 is face-centered cubic ($MgCu_2$-type) with $a = 7.766$ Å[27]. The compound UAl_3 is body-centered cubic with $a = 4.287$ Å and UAl_4 is orthorhombic with $a = 4.41$ Å, $b = 6.27$ Å and $c = 13.71$ Å[28,29]. The chemical composition of the UAl_4 phase and its density do not wholly accord with the formula UAl_4; the Al/U ratio ranges from 4.9 to 4.5 and the density is 5.7, whereas the X-ray density for UAl_4 is 6.5.

The reaction between uranium and aluminium begins at about 250 °C and becomes very rapid at 400 °C. Aluminium is virtually insoluble in α-uranium, the β-phase dissolves up to 0.7 at. % and the γ-phase can accomodate about 3 at. % Al in solid solution.

UAl_2 is only important as an impurity and to the extent to which it may alter the mechanical properties. Thus, a fine distribution of UAl_2 particles in an α-phase matrix may be obtained (< 3 at. % Al) by heat treatment, to give a textureless α-phase. This material has optimum mechanical properties and strength values.

High-aluminium alloys of uranium are made by adding uranium in the form of the oxide or fluoride to the molten aluminium, which is initially at 800 °C. Large ingots of these alloys are melted under a cryolite flux and cast in steel moulds.

The thermodynamic properties of the phases in the aluminium–uranium system have been reported upon by Johnson[10].

Gallium forms three compounds with uranium, UGa, UGa_2 and UGa_3. Their behaviour on vaporization has been studied by Alcock *et al.*[30] using the Knudsen effusion technique. The authors also investigated the *indium*–uranium system and confirmed previous findings[29] that only one compound, UIn_3, occurs in the system.

The thermodynamics of the indium rich part of the system has been described[31]. The cubic compounds UGa_3(a = 4.248 Å) and UIn_3(a = 4.601 Å) are isomorphous with UAl_3($AuCu_3$-type).

Uranium and metals of Group IV

Uranium forms a number of intermetallic compounds with the Group IV elements, among which are an interesting group of compounds with the general formula UX_3, and all having the Cu_3Au (Pm3m) structure. A theoretical consideration of the measured stabilities in these compounds has been given by Alcock *et al.*[30], who concluded that a covalent-bond model is a more successful approach to their structure than the ionic model.

Germanium forms with uranium a number of compounds: U_5Ge_3, UGe, U_3Ge_5, UGe_2 and UGe_3. The partial pressures over the different alloy phases have been used to obtain free-energy equations for each phase[6]. The compound UGe_3 is body-centered cubic with $a = 4.206$ Å.

The phase diagram of the uranium-*tin* system has not been completely investigated. The compounds U_3Sn_2, U_3Sn_5 and USn_3 have been found and the partial pressure of tin over these phases has been measured by means of the Knudsen effusion technique[6]. In addition EMF-measurements have been used to determine the free energy of formation of USn_3[31]. The results have been summarized by Johnson[10].

Lead forms with uranium the compounds UPb and UPb_3. The melting point of UPb_3 (1220 °C) is far above the value obtained by connecting the melting points of the two metals with a straight line, indicating a strong affinity of the two metals. For this reason a largely ionic bonding in UPb_3 has been assumed by Hume-Rothery[32]. Thermodynamic data on this system have been reviewed by Johnson[10].

Metals of the titanium, zirconium and hafnium subgroup form alloys with uranium that are of great technological interest because of their high solubility in the γ-phase of uranium.

The *titanium* alloys can be made by arc-melting pure uranium with pure titanium. The addition of titanium to uranium increases its hardness, strength and resistance to oxidation. The phase diagram of the system[33] is shown in Fig. 9. A continuous series of solid solutions is formed between the γ-modification of uranium and the β-modification of titanium. Moreover, the addition of titanium lowers the β–γ transformation from 771 °C to a eutectoid with the intermediate phase U_2Ti at 723 °C and 4 at. % Ti. The β-uranium modification dissolves about 1.5 at. % titanium at the eutectic temperature.

The compound U_2Ti is hexagonal with $a = 4.828$ Å and $c = 2.847$ Å[34]; it separates at 898 °C from the γ-solid solution, but only extends over a small range of homogeneity of about 0.5 at. %. A eutectic is formed between U_2Ti and α-titanium at

655°C and 83 at. % Ti. The solubility of uranium in α-titanium at the eutectic temperature is about 0.8 at. %.

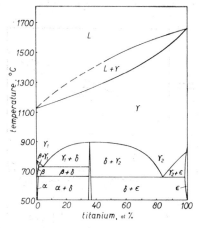

Fig. 9. Phase diagram of the titanium–uranium system.

The *zirconium*–uranium system (Fig. 10) is another example in which a complete range of solubility in the γ-phase occurs. The region is around 52.8 at. % zirconium. The addition of zirconium lowers the β–γ transition to a monotectoid at 693°C and 14.5 at. % Zr. At this temperature a $(\gamma_1 + \gamma_2)$ miscibility gap exists from 14.5 to 57 at. % zirconium[35]. Oxygen impurities enlarge the miscibility gap at both axes especially on the zirconium-rich boundary. The maximum in the gap occurs at 740°C and about 37 at. % zirconium.

The δ-phase, UZr_2, was formerly not observed. Whether it is formed or not depends on the method of making the alloy and on its purity[36]. Oxygen tends to prevent its formation.

The solubility of zirconium in α-uranium is 0.52 at. % at 662°C and in β-uranium 1.1 at. % at 693°C, the temperature at which the γ_1-phase decomposes. The effect of cooling rate on the decomposition of the γ-phase has been reported upon recently[37]. The phases formed, their distribution and hardness have been examined as a function of composition and cooling rate.

Hafnium is closely similar to zirconium in its properties and probably similar in its miscibility to γ-uranium. The δ-phase has, if stable at all, a narrow low-temperature region of stability. Of course hafnium is, because of its high absorption cross section to thermal neutrons, unsuitable in alloys for use in nuclear reactors.

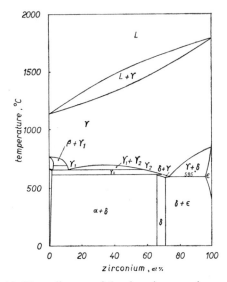

Fig. 10. Phase diagram of the zirconium–uranium system.

Uranium and metals of Group V

Vanadium forms alloys with uranium that melt at relatively low temperatures. The phase diagram is essentially similar to that of the chromium–uranium system (below). The maximum solubility of vanadium in γ-uranium at the eutectic temperature is 12 at. %.

The *niobium*–uranium system has been investigated intensively in recent years because of its interest in nuclear technology. Niobium gives an excellent stability to the γ-phase when added to the amount of approximately 22 at. % and also provides an increase in corrosion resistance. The various metastable phases α', α'' and $\gamma°$ and their mode of formation (similar to that in the uranium–molybdenum system) have been reported[38].

Tantalum–uranium alloys have been made by arc melting.

Addition of 5–11 at. % Ta to uranium appears to confer optimum resistance to corrosion in water at 100°C. The phase diagram of the system is shown in Fig. 11.

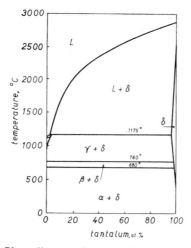

Fig. 11. Phase diagram of the tantalum–uranium system.

Uranium and metals of Group VI

In the *chromium*–uranium system, the eutectic temperature is 859°C and at 19.4 at. % chromium. The maximum solubility of chromium in the γ-phase is 4 at. % at the eutectic temperature[39]. The solubility in β-uranium is 1 at. % chromium, which is sufficient to retain the β-phase on quenching, but it is transformed during cold deformation. Suitable amounts of chromium have a grain refining action on β-quenched, hot-rolled uranium and thus improve the thermal-cycling stability of the uranium. The optimum addition is 0.78 at. %. Uranium–chromium alloys can be made by co-reduction of UF_4 with chromium oxide; melting of the alloys is feasible in graphite crucibles.

The phase diagram of the *molybdenum*–uranium system[40,41] is shown in Fig. 12. The alloys of uranium with molybdenum are very important as nuclear fuels, particularly in naval reactors and in the pressurized water reactor (PWR) in Shippingport (USA).

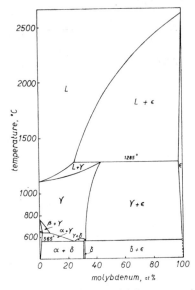

Fig. 12. Phase diagram of the molybdenum–uranium system.

The chemical and thermal stability of molybdenum metal and its low neutron absorption cross section make it of great value in nuclear reactors. The solubility of molybdenum in the α-phase uranium is less than 0.5 at. % (at 640 °C), whereas the solubility in the β-phase is 2.2 at. %. The maximum solubility of molybdenum in the γ-phase uranium is 42 at. % at the peritectic temperature of 1285 °C. The chief interest is in the γ-structure obtained by quenching. The γ-phase was once thought to be stabilized at room temperature in the vicinity of 25 to 34 at. % molybdenum, but it is now known to be metastable in this range, below a temperature of 565 °C[42]. The critical alloying addition of molybdenum that will just retain the γ-structure in uranium on quenching in water is 15 at. %. The alloy must be homogenized at 900 °C for seven days and then water quenched.

The 31 at. % alloy is transformed to the δ-form on prolonged annealing at 550 °C, just below the limit of stability of the γ-phase[43]. Alloys with a low molybdenum content, e.g. 1.6 at. % molybdenum, are transformed to $\alpha + \gamma'$ phases of uranium below 565 °C. The isothermal transformation of such alloys has been studied by Östberg et al.[44]. The metallurgical aspects of the molybdenum–uranium system have been discussed at length by Wilkinson[1].

Uranium and metals of Group VII

The rhenium–uranium system appears to be very similar to the uranium–molybdenum system. The difference is mainly the absence of the acicular α'-structure of uranium at low rhenium contents[45,46]. The α- and β-phases of uranium can be retained at room temperature by the addition of rhenium and rapid cooling. Rhenium, although hexagonal at room temperature, is miscible with γ-uranium to an extent of 16 at. %; the alloy thus formed is retained in the γ-phase on water quenching, whereas with 3.5 and 8.5 at. % rhenium metastable modifications of the α-structure (α' and α'') are obtained in rapidly cooled γ-alloys. The γ-phase is thus transformed to α-phase without an intermediary β-phase stage.

The phase URe_2 is formed when the proportion is over 10 at. % rhenium; from this the δ-phase, U_2Re, is produced at 700 °C by the peritectoid reaction[47]:

$$\gamma\text{-}U + URe_2 \xrightarrow{700\ °C} U_2Re$$

Uranium and metals of Group VIII

From the iron group metals, the iron–uranium system is of particular interest since uranium containing small additions of iron can be grain-refined by water quenching from 750 °C. Recent work[48] has indicated that most of the iron is in solid solution in "adjusted" uranium, even in the α-range. The solubility of iron in solid uranium has been determined up to a concentration of 0.3 at. %[49].

At 660 °C α-uranium can dissolve 0.003 wt. % (0.013 at. %) iron and at the eutectic temperature for the β-phase (661 °C), 0.12 at. %. The maximum solubility in the β-phase is 0.37 at. % at 763 °C and at the γ-eutectic point, 0.65 at. %. In the γ-phase the solubility of iron increases with increasing temperature, being 1.33 at. % at 800 °C.

The compound U_6Fe is formed by a peritectic reaction between the γ-solid solution and the liquid; it has a body-centered tetragonal lattice, with $a = 10.31$ Å and $c = 5.24$ Å[26]. The other compound in the iron–uranium system, UFe_2, is face-centered cubic with $a = 7.042$ Å[50].

With the metals of the *platinum group* uranium forms compounds of the type UX_3, of which UPd_3 and UPt_3 are hexagonal and URu_3 is face-centered cubic. In addition the compounds UPt_2 and UIr_2 are known[51].

The alloys of *ruthenium* with uranium are mainly of interest because ruthenium is a product of fission found in fissium (see p. 53). The maximum solubility of ruthenium in γ-uranium is 7.5 at. % (886 °C), but in α- and β-uranium less than 2 at. % ruthenium can be dissolved. The metastable phases α' and $\gamma°$ produced by fast γ-quenching of uranium alloys containing up to 10 at. % ruthenium have been examined and the absence of an α'' phase is reported[52]. The eutectic temperature is 886 °C (at 18.5 at. % Ru). A large number of intermetallic compounds occur in the ruthenium–uranium system: U_2Ru, URu, U_3Ru_4, U_3Ru_5 and URu_3 are formed peritectically at 937 °C, 1158 °C, 1163 °C, 1182 °C and 1850 °C respectively[53]. The URu phase melts congruently at 1158 °C.

The uranium–plutonium system

In the plutonium–uranium system the phases UPu and UPu_3 have been observed; both have an extensive range of homogeneity. The solubility of plutonium in both α- and β-uranium is rather extensive being 18 at. % in β at 600 °C and 15 at. % in α at 560 °C[38] falling to 12 at. % at room temperature. When the solid solubility at room temperature is exceeded, the δ-phase, the only room temperature intermediate, appears[54]. Of the plutonium alloys, those with 15 at. % have especially poor dimensional stability on irradiation. Better stability is shown by material with 20 % plutonium, particularly when it has been extruded.

The cast $(\alpha + \delta)$ alloy containing 20 % plutonium is too brittle to be removed from the mould without cracking. Moreover, it is pyrophoric and it is extremely susceptible to radiation damage. It thus appears that suitable additions of molybdenum (5 wt. %) will stabilize the cubic γ-phase[54]. The lattice parameter of the alloy is 3.42 Å. A still further improvement is obtained by addition of 10 wt.% of fissium. It is important that U–Pu–Zr alloys are compatible with 304 stainless steel and do not form molten reaction products when heated to temperatures above $800\,°C$[55]. The increasing of the solidus temperature of uranium–plutonium alloys is very important for fuel applications.

Uranium and fissium

Fissium is the equilibrium concentration of fission-product elements left by the pyrometallurgical reprocessing cycle. When spent uranium–plutonium fuel is melted in a zirconia crucible and kept molten for several hours at temperatures between 1300 and $1400\,°C$, the fission products xenon, krypton and caesium are removed by volatilization and the rare earths barium and strontium react with the crucible. The fission products that remain in the regulus are molybdenum, ruthenium, rhodium, palladium, zirconium, niobium and tellurium. These elements thus collected have been grouped together and called "fissium" (Fs), a term which first appeared in 1955. Considered as an alloying additive, it has the composition 50 wt. % Mo, 36 wt. % Ru, 10 wt. % Pd, 6 wt. % Rh, 4 wt. % Zr and 0.2 wt. % Nb. The niobium content is usually ignored. The phase relationships and physical properties of simulated alloys are assumed to be not very greatly different from those in the actual fission-produced alloys[56].

The uranium–5 wt. % Fs alloy melts at a lower temperature than uranium; the liquidus and solidus temperatures are reported[57] to be 1081 °C and 1002 °C respectively. The density is 18.0 (25 °C) and 17.3 (800 °C). The γ–β transformation in uranium is suppressed by increasing amounts of fissium. In 3 wt. % Fs it is 682 °C, instead of 769.4 °C for unalloyed uranium. When the temperature falls to 660 °C, $\beta + U_2Ru$ begins to be coprecipitated. In the 5 wt. % Fs-alloy, U_2Ru is precipitated at 725 °C when the γ-phase is slowly cooled. Heat capacity studies of uranium–fissium alloys have been reported[58]. The U–Pu–Fs alloy contains seven components and as many as five phases have been detected in equilibrium[59]. Transformations in this system are extremely complex; the γ-phase was found to be retained only in alloys with more than 10 wt. % fissium.

Uranium alloys in nuclear technology

The marked tendency towards the formation by uranium of intermetallic compounds with a variety of alloying metals and the absence of extensive ranges of solid solution of these metals in the unique α-uranium and β-uranium structures is an outstanding feature of uranium alloy systems. As already indicated, the primary aim in alloying uranium is to improve its dimensional stability and corrosion resistance. The approach to the use of alloying has been in two principal directions: (a) to stabilize the isotropic γ-phase, (b) to modify the kinetics of the β and γ transition so as to yield a structure combining random orientation and fine grain size ("α-alloy systems").

γ-Uranium miscible systems

The elements zirconium, molybdenum, titanium, niobium and ruthenium are very miscible in γ-uranium and, moreover, the γ-phase may be retained at room temperature although it is not the thermodynamically stable state. These alloys also display a variety of metastable phases of great technological interest. Furthermore, a miscibility gap, containing two different γ-compositions ($\gamma_1 + \gamma_2$) is characteristic for the systems U–Ti, U–Zr and U–Nb. In the uranium–zirconium system, for instance, the miscibility gap extends from 14.5 to 57 at. % Zr over a temperature range from 693 °C to 735 °C. In the ternary systems, for example that with titanium, the smaller titanium atom reduces the strain energy in the γ-solid solution and hence titanium causes the disappearance of the gap[60] when added to alloys in the uranium–zirconium system.

The presence of high-temperature phases and metastable transition phases in the binary uranium alloy systems, in which the γ-miscibility exceeds a few at. %, has now been well documented[33, 38, 42, 52]. The stability of the γ-phase may be increased further by the addition of a third element. For instance, the effect of Cr, Nb, Re, Ru and Zr respectively on uranium–molybdenum alloys has been described by Cabana and Donzé[61].

It is interesting that the γ-miscible elements tend to form at least one intermediate subgamma phase, labelled δ, which is generally somewhat metallic in character.

α-Alloy systems

A disadvantage of the highly miscible γ-uranium systems is that large amounts of neutron absorbent materials are introduced into the reactor fuel, especially when natural or slightly enriched uranium are being used. Parasitic neutron absorption by the added elements reduces the advantages of this kind of alloying. Because of this, quite small additions are made to the uranium to control its grain size and grain orientation. Uranium to which these small additions have been made retains the α-structure, but at the same time the growth and swelling are minimized and the alloying in combination with an appropriate heat treatment (β-quenching) causes the formation of a large number of small crystallites which are randomly oriented. Indeed these small additions to uranium make possible what once was thought to be impossible, namely the thermal cycling of uranium through the α–β transformation temperature without undue deformation. The elements most effective in preventing deformation during these operations are in decreasing order of effectiveness, Pt, Nb, Cr, Ru, Si, Mo and Zr. Thus, the addition of 1.2 wt. % Mo appears to be more effective than 7 wt.% Zr. Moreover, ternary systems are particularly successful, such a one is U–2 wt.% Mo–0.5 wt.% Zr[62].

Irradiation swelling is a matter of concern to fuel designers. Serious swelling of the fuel occurs when the gas bubbles, caused by the rising concentration of xenon and krypton produced as fission proceeds, increase in size. The presence in uranium of some elements, especially aluminium, has a marked effect on the swelling behaviour. In "adjusted" uranium minute additions of aluminium (400–1200 ppm), iron (300 ppm) and carbon (600 ppm) form fine precipitates of UAl_2 and U_6Fe at grain boundaries after an appropriate heat treatment followed by β-quenching and annealing in the α-phase. Iron alone is sufficient to produce grain refinement, but addition of aluminium is necessary to inhibit cracking which tends to occur as a result of severe quenching. The major advantage of these small additions is, however, to reduce swelling, particularly between 450° and 600 °C. The precise way in which these additions promote swelling resistance is not yet clear. It has been suggested that the precipitates act as nucleation centres and thus hinder migration of the gas bubbles and their subsequent coalescence to larger ones[63]. Recent work by Hudson[64] does not, however, support some of the general ideas on swelling resistance.

REFERENCES

1 W. D. WILKINSON, *Uranium Metallurgy*, Vol. II, *Uranium Corrosion and Alloys*, Interscience Publishers, New York, London (1962).
2 J. H. GITTUS, *Uranium* (series: *Metallurgy of the Rarer Metals*), Butterworths, London (1963).
3 H. A. WILHELM, *Trans. Am. Soc. Metals*, 42 (1950), 1311.
4 R. W. BUZZARD, *J. Res. Nat. But. Standards*, 52 (1954) 149.
5 R. W. BUZZARD, *J. Res. Nat. Bur. Standards*, 53 (1959) 291.
6 C. B. ALCOCK AND P. GRIEVESON, *J. Inst. Metals*, 90 (1961) 304.
7 M. I. IVANOV AND V. A. TUMBAKOV, *Atomnaya Energiya*, 7 (1959) 33.
8 P. CHIOTTI, *Trans. Am. Inst. Min. Eng.*, 206 (1958) 562.
9 P. CHIOTTI AND G. R. KILP, *Trans. AIME*, 218 (1960) 41.
10 I. JOHNSON, *Proceedings Symposium on Compounds of Interest*, *AIME* Nucl. Met. X. (1964), p. 171.

11   A. E. MARTIN, I. JOHNSON AND H. M. FEDER, *Trans. Met. Soc. AIME*, 221 (1961) 789.

12   I. JOHNSON AND H. M. FEDER, *Trans. Met. Soc. AIME*, 224 (1962), 468.

13   B. R. T. FROST, *J. Inst. Metals*, 82 (1953) 456.

14   H. C. FORSBERG, *ORNL-Report* 2885 (1960).

15   F. BERTAUT AND P. BLUM, *Compt. Rend.*, 229 (1949) 666.

16   A. ZALKIN AND D. H. TEMPLETON, *Acta Cryst.*, 6 (1953) 269.

17   A. ZALKIN AND D. H. TEMPLETON, *J. Chem. Phys.*, 18 (1950) 391.

18   G. BECKMANN AND R. KIESSLING, *Nature*, 178 (1956) 1341.

19   P. BLUM AND F. BERTAUT, *Acta Cryst.*, 7 (1954) 81.

20   J. L. ANDRIEUX AND P. BLUM, *Compt. Rend.*, 229 (1949) 210.

21   B. W. HOWLETT, *J. Inst. Met.*, 88 (1959/60) 91.

22   J. L. ANDRIEUX, *Thesis*, Paris (1929); see also *Ann. Chim.*, 10 (1929) 423.

23   G. VERSTEEG, private communication.

24   C. B. ALCOCK AND P. GRIEVESON, *Thermodynamics of Nuclear Materials*, Proceedings of a Symposium (1962), IAEA, Vienna (1963), p. 563.

25   W. M. ALBRECHT AND B. G. KOEHL, *Proceedings of the Second International Conference on the Peaceful Uses of Atomic Energy, Geneva (1958)*, United Nations (1959) Vol. 6, p. 116.

26   M. D. JEPSON, *ibid.*, p. 42.

27   G. KATZ AND A. J. JACOBS, *J. Nucl. Mat.*, 5 (1962) 338.

28   R. E. RUNDLE, *Acta Cryst.*, 2 (1949) 148.

29   B. R. T. FROST AND J. T. MASKREY, *J. Inst. Metals*, 82 (1953) 171.

30   C. B. ALCOCK, J. B. CORNISH AND P. GRIEVESON, *Thermodynamics, Proceedings of a Symposium* (1965), IAEA, Vienna, Vol. I (1966), p. 211.

31   I. JOHNSON AND H. M. FEDER, *Thermodynamics of Nuclear Materials*, Proceedings of a Symposium (1962), IAEA, Vienna (1963), p. 319.

32   H. HUME-ROTHERY, *J. Inst. Metals*, 83 (1955) 535.

33   A. G. KNAPTON, *J. Inst. Metals*, 83 (1955) 1628.

34   D. J. MURPHY, *Trans. Amer. Soc. Metals*, 50 (1958) 884.

35   H. H. CHISWIK, *Proceedings of the Second International Conference on the Peaceful Uses of Atomic Energy, Geneva* (1958), United Nations (1959), Vol. 6, p. 394.

36   F. A. ROUGH, A. E. AUSTIN, A. A. BAUER AND J. R. DOIG, *BMI-Report* 1092 (1956).

37   R. F. HILLS, B. R. BUTCHER, B. W. HOWLETT AND D. STEWART, *AERE-R-Report* 4557 (1964).

38   K. TANGRI AND D. K. CHAUDURA, *J. Nucl. Mat.*, 15 (1965) 278.

39   A. H. DAANE AND A. S. WILSON, *ISC-Report* 564 (1955).

40   A. E. DWIGHT, *J. Nucl. Mat.*, 2 (1960) 81.

41   J. LEHMANN, *Thesis*, Paris (1959).

42   K. TANGRI AND G. J. WILLIAMS, *J. Nucl. Mat.*, 4 (1961) 226.

43   R. SIFFERLEN, J. GUILLAUMIN AND A. SAULNIER, *J. Nucl. Mat.*, 23 (1967) 270.

44   G. OSTBERG AND B. LEHTINEN, *J. Nucl. Mat.*, 23 (1967) 123.

45   R. J. JACKSON, D. WILLIAMS AND W. L. LARSEN, *J. Less-Common Metals*, 5 (1963) 443.

46   R. J. JACKSON AND W. L. LARSEN, *J. Nucl. Mat.*, 21 (1967) 277.

47   R. J. JACKSON AND W. L. LARSEN, *J. Nucl. Mat.*, 21 (1967) 282.
48   A. F. SMITH, *J. Less-Common Metals*, 9 (1965) 233.
49   N. SWINDELLS, *J. Nucl. Mat.*, 18 (1966) 261.
50   J. D. GROGAN, *J. Inst. Metals*, 77 (1950) 571.
51   T. J. HEAL AND G. J. WILLIAMS, *Acta Cryst.*, 8 (1955) 494.
52   K. TANGRI, D. K. CHAUDURA AND C. N. RAO, *J. Nucl. Mat.*, 15 (1965) 288.
53   J. J. PARK, *J. Res. Nat. Bur. Stand.*, 72A, (1968) 1.
54   K. F. SMITH AND L. R. KELMAN, *ANL-Report* 5677 (1957).
55   R. E. MACHEREY, L. R. KELMAN AND J. H. KITTEL, *ANL-Report* 7120 (1965).
56   M. V. NEVITT AND S. T. ZEGLER, *J. Nucl. Mat.*, 1 (1959) 6.
57   H. A. SALLER, R. F. DICKERSON, A. A. BAUER AND N. E. DANIEL, *BMI-Report* 1123 (1956).
58   H. SAVAGE, *J. Nucl. Mat.*, 25 (1968) 249.
59   O. L. KRUGER, *J. Nucl. Mat.*, 19 (1966) 29.
60   B. W. HOWLETT, *J. Nucl. Mat.*, 1 (1959) 289.
61   G. CABANA AND G. DONZÉ, *J. Nucl. Mat.*, 4 (1959) 364.
62   A. B. MCINTOSH AND T. J. HEAL, *Proceedings of the Second International Conference on the Peaceful Uses of Atomic Energy, Geneva (1958)*, United Nations (1959), Vol. 6, p. 413.
63   R. S. BARNES, R. G. BELLAMY, B. R. BUTCHER AND P. G. MARDON, *Proceedings of the Third International Conference on the Peaceful Uses of Atomic Energy, Geneva (1964)*, United Nations (1965), Vol. 11, p. 218.
64   B. HUDSON, *J. Nucl. Mat.*, 22 (1967) 121.

CHAPTER 5

Uranium Hydride

Uranium hydride, UH_3, is an interesting and important substance, since it is the intermediate in the preparation on a laboratory scale of most uranium compounds for which uranium is the starting material. The formation of UH_3 and its decomposition at higher temperatures desintegrate massive uranium metal, leaving it finely divided and thus in a form which is excellently suitable for the synthesis of uranium compounds. This is certainly the most important use of the hydride.

Crystallographic and magnetic properties

Two polymorphs of uranium hydride are now known to exist of which β-UH_3 is the form commonly encountered. It was found from X-ray data[1] that it has a primitive cubic crystal lattice (β-W type), with $a = 6.6310$ Å. The density is 10.92. The structure of the isomorphous deuteride, UD_3, which is more easy to study by means of neutron diffraction, has been examined[1,2] and the deuterium atoms have been shown to lie in distorted tetrahedra, equidistant from four uranium atoms, with a U–D distance of 2.32 Å.

At low temperatures there is another cubic form of UH_3[3] with $a = 4.160$ Å. The cell of this α-form contains two molecules of UH_3. The α-UH_3 structure is stable with respect to β-UH_3 up to about 100 °C, but the reverse transformation does not seem to take place at a measurable rate.

A ferromagnetic transition in UH_3 at 173 °K has been observed[4].

Preparation

Uranium metal in the mass reacts rapidly with hydrogen at about 250 °C to form the hydride UH_3. The reaction results in a complete desintegration ot the metallic structure, the product being a very voluminous black powder with a bulk density of 3.4.

The reaction velocity has a maximum value at 225 °C. Above 250 °C the reaction rate falls off. At temperatures above 400 °C dissociation of UH_3 begins; and the hydrogen pressure reaches a value of 1 atmosphere at about 435 °C.

Finely divided uranium powder obtained by decomposition of the hydride reacts with hydrogen at much lower temperatures than the massive metal, reaction taking place at 0 °C; even at -80 °C some reaction has been observed[5].

Thermodynamic properties and the phase diagram

The hydrogen pressure of the equilibrium:

$$UH_3 \rightleftarrows U + \tfrac{3}{2} H_2$$

between 450 °C and 650 °C has been measured as a function of the temperature by Libowitz and Gibb[6], who obtained:

$$\log p_{H_2}(\text{atm}) = \frac{-4410}{T} + 6.26$$

The heat of reaction, calculated from these measurements, is 30.3 kcal/mole and is in good agreement with the heat of formation of UH_3 (30.35 kcal/mole) found from calorimetric measurements[7]. The measurements of the hydrogen pressures by Libowitz and Gibb, have been extended by Chevallier et al.[8] to temperatures of 850 °C and hydrogen pressures of 150 atm. Their results are in reasonable agreement with those previously mentioned and may be represented by

$$\log p_{H_2}(\text{atm}) = \frac{-4200}{T} + 6.05$$

For the entropy of UH_3 the value $S_{298}^0 = 15.27$ cal/deg.mole has been found from low-temperature heat-capacity measurements[9]; at 298 °K the value of $c_p = 11.78$ cal/deg.mole.

The phase diagram of the uranium–hydrogen system is shown in Fig. 13. The solubility of hydrogen in uranium has been measured by Saller and Rough[10], Libowitz and Gibb[6] and at high temperatures by Chevallier *et al.*[8]. The solid solubility increases considerably with increasing temperature up to 665 °C; at this temperature (the α–β transition temperature in uranium) solubility has reached the maximum value, corresponding to the atomic ratio H/U = 1.1. At higher temperatures the solubility of hydrogen in β- and γ-uranium decreases.

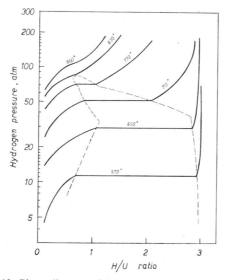

Fig. 13. Phase diagram of the uranium–hydrogen system.

At the hydrogen-rich portion of the phase diagram, the compound UH_3 does not deviate from stoichiometry to any measurable degree at room temperature but does so to a significant degree at elevated temperatures[6, 8]. For example, at 650 °C the H/U ratio is 2.85, decreasing rapidly at higher temperatures. Calculations by Libowitz[11] showed that the deviations from stoichiometry in UH_3

are due to the formation of hydrogen vacancies in the lattice, the formula thus being UH_{3-x}.

At 820 °C and about 83 atm, the composition of the solid solution and the non-stoichiometric hydride UH_{3-x} become equal (H/U = 0.70). Above this temperature the system forms only a continuous solution (Fig. 13).

Chemical properties

The compound UH_3 is a highly reactive substance. It is pyrophoric and undergoes reactions with a great variety of materials. This reactivity forms the basis of the preparation of a large number of uranium compounds. Thus, nitrogen reacts above 250 °C to give the nitride U_2N_3 and H_2S gives the sulphide US_2. The gases CO_2 and H_2O react above 250 °C to give the oxide UO_2, whereas reactions with the halogens give the tetrahalides. Other reactions have been reviewed extensively by Katz and Rabinowitch[12]. It is interesting to note that uranium hydride loses its pyrophoric character when it is treated with hydrogen at high pressures[8], probably as a result of particle growth.

REFERENCES

1   R. E. RUNDLE, *J. Am. Chem. Soc.*, 69 (1947) 1719.
2   R. E. RUNDLE, *J. Am. Chem. Soc.*, 73 (1951) 4172.
3   R. N. R. MULFORD, F. H. ELLINGER AND W. H. ZACHARIASEN, *J. Am. Chem. Soc.*, 76 (1954) 297.
4   D. M. GRUEN, *J. Chem. Phys.*, 23 (1955) 1708.
5   J. J. KATZ AND E. RABINOWITCH, *The Chemistry of Uranium*, Part I, McGraw-Hill Book Company, Inc., New York (1951), p. 194.
6   G. G. LIBOWITZ AND T. R. P. GIBB, Jr., *J. Phys. Chem.*, 61 (1957) 793.
7   B. M. ABRAHAM AND H. E. FLOTOW, *J. Am. Chem. Soc.*, 77 (1955) 1446.
8   J. CHEVALLIER, P. DESRE AND J. SPITZ, *J. Nucl. Mat.*, 23 (1967) 289.
9   H. E. FLOTOW, H. R. LOHR, B. M. ABRAHAM AND D. W. OSBORNE, *J. Am. Chem. Soc.*, 81 (1959) 3529.
10 H. A. SALLER AND F. A. ROUGH, *BMI-Report* 1000 (1955).
11 G. G. LIBOWITZ, *J. Appl. Phys.*, 33 (1962) 399.
12 J. J. KATZ AND E. RABINOWITCH, *The Chemistry of Uranium*, Part I, McGraw-Hill Book Company, Inc., New York (1951), p. 201.

CHAPTER 6

The Oxides of Uranium

Introduction

The uranium–oxygen system is one of the most complicated of the binary systems. Thus is not only due to the existence of a large number of oxide phases, but also to the fact that deviations of stoichiometry are more the rule than the exception. Of the various phases in the system, uranium dioxide, UO_2, has particular interest in nuclear technology since it is the nuclear fuel now most widely used in energy-producing reactors. Numerous investigations on the uranium–oxygen system have been and are still being published. An excellent and extensive review, including the various aspects of technological work, but covering the literature only up to 1961, has been given in *Uranium dioxide*, edited by J. Belle[1]. A more recent, but specialized review has been published by the IAEA in 1965, entitled *Thermodynamic and transport properties of uranium dioxide and related phases*[2].

Since these reviews present so many intercomparisons of important information, many references have been made to them in this chapter, rather than to the original papers. In addition, some even more recent work has been incorporated.

The phase diagram

Many investigators have examined the phase diagram of the uranium–oxygen system, using a variety of techniques. The existence of at least four thermodynamically stable oxide phases, UO_2, U_4O_9, U_3O_8 and UO_3, have now been well established.

In addition to these several metastable phases, such as the U_3O_7 and U_2O_5[3] phases, have been reported. At this point only a general description of the phase diagram will be given, and later in the chapter the preparation and the physicochemical properties of the various phases mentioned will be dealt with.

The phase diagram at moderate temperatures ($< 1500\,°C$)

The solubility of oxygen in uranium metal is low, increasing from 0.05 at.% at the melting point ($1132\,°C$) to 0.4 at.% at $2000\,°C$[3a]. The oxide in equilibrium with uranium is UO_2. Although the preparation of a lower oxide phase, UO, has been reported its existence as a pure, stable phase could not be proved; attempts to prepare it by direct reaction between uranium and uranium dioxide have always failed. There are strong indications that the "UO"-phase exists only because of the presence of such impurities as carbon or nitrogen[4,5]. The lattice parameter of the cubic monoxide solid solution, U(O, C, N), which is of NaCl-type, is $4.92\,Å$[4].

By far the most important phase in the system is the oxide UO_2. It is now generally agreed that the UO_2 phase does not extend below the composition $UO_{2.00}$ at moderate temperatures (say below $1000\,°C$). At room temperature, and up to at least $300\,°C$ oxygen fails to enter the UO_2 structure to form stable solid solutions. Nevertheless, at higher temperatures oxygen does penetrate the UO_2 lattice interstitially to give material of the composition UO_{2+x}, in which the value of x depends on the temperature, the surface area of the sample and the partial pressure of oxygen. The limiting value of x, at which UO_{2+x} is in equilibrium with the U_4O_9 phase, increases with temperature to a value of 0.17 at $950\,°C$ and 0.244 at $1123\,°C$, the highest temperature at which the U_4O_9 phase can exist[6]. Above that temperature the limiting value of x, given by the equilibrium between UO_{2+x} and U_3O_{8-z}, increases gradually.

The tetragonal U_4O_9 phase has a homogeneity range, the extent of which has not been well defined at low temperatures. The upper limit for the phase is O/U $= 2.25$ at $1123\,°C$; at this temperature there is an (invariant) quadruple point at which the phases UO_{2+x},

U_4O_{9-y}, U_3O_{8-z} and O_2 are in equilibrium. Recently, the existence of three different U_4O_9 phases has been shown; their relationships will be discussed below (see p. 80).

The U_3O_8 phase has also a range of homogeneity, varying from $O/U = 2.667$ at room temperature to 2.61 at the quadruple point (1123 °C). The oxide UO_3 is the thermodynamically stable oxide phase which is in equilibrium with oxygen under atmospheric pressure at temperatures below 500 °C. It exists in at least six different crystallographic modifications of which the yellow γ-UO_3 is the stable one.

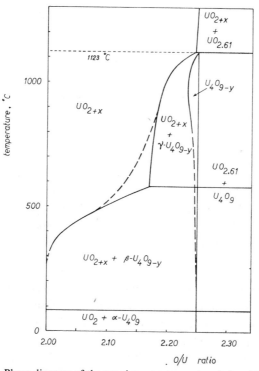

Fig. 14. Phase diagram of the uranium–oxygen system below 1500 °C.

In addition to the stable oxide phases just described, there is a metastable phase, approximating U_3O_7 in composition, which

TABLE 6

PROPERTIES OF URANIUM OXIDE PHASES*

Oxide phase	Structure	$-\Delta H_{298}$ (kcal/mole)	$S°_{298}$ (cal/deg·mole)
UO_2	cubic, $a = 5.470$ Å	259.2	18.41
α-U_4O_9	cubic, $a = 21.77$ Å	1078	83.53
α-U_3O_7 ⎱** β-U_3O_7 ⎰	tetragonal, $a = 5.46$, $c/a = 0.99$ or 1.01 $a = 5.371$, $c/a = 1.03$	818.4	51.09 51.51
α-U_3O_8	orthorhombic, $a = 6.72$ Å, $b = 11.96$ Å, $c = 4.15$ Å	854.1	67.5
γ-UO_3	monoclinic, pseudo-tetragonal, $a = b = 6.89$ Å, $c = 19.94$ Å, $\gamma = 90.34$	293.5	23.6

*References, see text.
**Metastable phases.

can be obtained by the oxidation of UO_2 below 200°C. It was recognized as a separate phase by Hering and Perio[7] and by Grønvold[8]. Later, Westrum and Grønvold[9] proved definitely the existence of two crystallographically different U_3O_7 phases; their characterization will be given below. When the U_3O_7 phase is annealed at temperatures above 600°C a rearrangement occurs:

$$2\,U_3O_7 \rightarrow U_4O_9 + 2\,UO_{2.61}$$

The preparation of an oxide, approximating U_2O_5 in composition, has been reported recently[3]. It is an unstable, hexagonal compound ($a = 3.996$ Å, $c = 4.117$ Å) which above 250°C disproportionates into U_4O_9 and U_3O_8.

Thus, the phase diagram from 500°C to 1500°C (Fig. 14) can now be regarded as fairly well established. But despite very many investigations, uncertainties in the phase diagram below 500°C, mainly with respect to the homogeneity range of UO_{2+x} and U_4O_{9-y} still remain.

The phase diagram at high temperatures (> 1500°C)

The use of UO_2 in high-temperature reactors has made an extensive study of its properties from $\sim$ 1500°C up to the melting point ($\sim$ 2800°C) necessary. It is now generally agreed that UO_2 becomes hypostoichiometric when it is heated in inert atmospheres above 1600°C. The location of the boundary line of the UO_{2-x} phase has been investigated by several authors[10,11,12] and the results are in fair agreement. A tentative phase diagram for the U–UO_2 part of the uranium–oxygen system is included in Fig. 15. From the figure it can be seen that the O/U ratio of the oxide gradually changes from 1.92 at 1800°C to 1.6 at 2470°C[10].

The most noticeable feature of the diagram is a wide liquid-miscibility gap. There is, however, disagreement concerning the top of this gap and the composition of the uranium-rich liquid[10,11]. The O/U ratio of the oxygen-rich liquid at the monotectic temperature of 2470°C is reported to be 1.18[13]. The best value for the melting point of UO_2 is 2800°C. Recent investigations have shown that the melting point of UO_2 has a maximum value at a slightly hypostoichiometric composition. The maximum reported is 2850°C

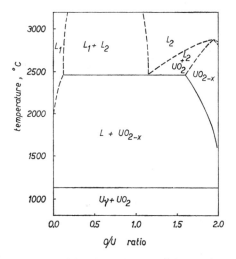

Fig. 15. The U–UO_2 part of the phase diagram of the uranium–oxygen system above 1500°C.

at the composition $UO_{1.95}$. In samples with a O/U ratio of less than 1.95 the temperature at which melting begins, decreases to values substantially less than for $UO_{2.00}$[14].

The uranium dioxide phase

Crystallographic properties; non-stoichiometry in UO_2

Uranium dioxide has a face-centered cubic structure (fluorite-type, space group Fm3m) with lattice parameter $a = 5.470$ Å at the stoichiometric composition[2]. The elementary cell is face-centered with respect to the uranium ions, with the oxygen ions at $\frac{1}{4}, \frac{1}{4}, \frac{1}{4}$ sites. The unit cell contains four molecules of UO_2 as well as four interstitial holes at $\frac{1}{2}, \frac{1}{2}, \frac{1}{2}$, equidistant from the eight oxygen atoms (Fig. 16). An apparent displacement of the oxygen atoms at 800 °C has been observed[15] which has been ascribed to anharmonic vibrations.

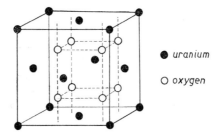

Fig. 16. Unit cell of UO_2.

When UO_2 is oxidized, oxygen is taken up in interstitial positions to form the UO_{2+x} phase which has the same space group as that for stoichiometric UO_2. That oxygen is accomodated interstitially follows from lattice, density and diffusion measurements, oxygen being the diffusing species[16]. The way in which oxygen is taken up has been examined by Willis[15] on single crystals of the composition $UO_{2.13}$ using neutron diffraction at high temperatures where the non-stoichiometric phase is stable. Willis found that when an extra oxygen enters the UO_2 lattice two different types of interstitial oxygens are created by the displacement of normal oxygen atoms from their sites, thus leaving normal-oxygen vacancies in the oxygen sublattice. The uranium atoms are assumed not to be displaced from their fluorite-type positions. The interstitial oxygen atoms thus do not occupy the large interstitial holes at the fourfold positions $\frac{1}{2}, \frac{1}{2}, \frac{1}{2}$; $\frac{1}{2}, 0, 0$; $0, \frac{1}{2}, 0$ and $0, 0, \frac{1}{2}$, but at positions ~ 1 Å from the centres of these interstices along the 110 and 111 directions (designated as O′ and O″ respectively) (Fig. 17). The composition of $UO_{2.13}$ can be represented, according to Willis, as $UO_{1.82}O'_{0.08}O''_{0.28}$.

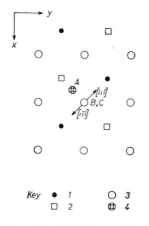

Fig. 17. Formation of UO_{2+x} (after Willis[15]).
The O′ atom at A ejects the two nearest oxygen atoms at B, C which are displaced along the $\langle 111 \rangle$ direction towards the adjacent holes, being converted thereby to interstitial O″ atoms.
key: 1 = uranium at $z = \frac{1}{4} a$
2 = hole at $z = \frac{1}{4} a$
3 = normal oxygen at $z = 0$
4 = O′ oxygen at $z = \frac{1}{4} a$

The lattice parameter of stoichiometric UO_2 at 20 °C is 5.470 Å which corresponds to a theoretical density of 10.952. This value is close to the density 10.950 measured for a single crystal of UO_2[2]. The unit cell of UO_2 contracts with increasing oxygen concentration. This contraction is undoubtedly due to the fact that the U^{5+} ion, formed upon oxidation, is smaller than the U^{4+} ion.

References p. 93

The density increases regularly with change in composition from 10.95 for $UO_{2.00}$ to 11.21 for U_4O_9, but the agreement between the various sets of measurements is not very good[2], mainly because of experimental difficulties in the measurement of precise lattice constants or densities, since: (a) the UO_{2+x} structure is not easily quenched-in owing to rapid oxygen diffusion, (b) the densities of powders which have been exposed to air are lower than would be expected because of the chemisorption of oxygen on the surface. Thus many $UO_{2.0}$ preparations have densities as low as 10.89. An obvious conclusion is that the lattice parameter of UO_{2+x} is not a suitable measure for the value of x.

Thermodynamic properties

The single-phase region. An extensive assessment of the data relating to the partly disordered UO_{2+x} single phase region at high temperatures has been given in the book by Rand and Kubaschewski[17]. For the equilibrium

$$O_2 \rightleftarrows 2[O]_{UO_{2+x}}$$

they listed smoothed values for the partial molar enthalpies and entropies expressed as a function of x and T. These had been obtained from measurements made by different techniques, such as those of EMF, of oxygen partial pressures and of equilibria with CO/CO_2 or H_2/H_2O. The assessment could not take into account more recent measurements[19,20,21,22]; and, although agreement in the values for the partial molar quantities is generally good for O/U ratios > 2.01, at lower O/U ratios it is much less satisfactory[2]. The disagreement mainly concerns the minimum in the $-\Delta \bar{S}_{O_2}$ curve at a point close to the stoichiometric composition. This minimum cannot be explained by using a statistical-mechanical model[23] which applies the configurational entropy of interstitial ions and the associated small temperature effect, as it can for the observed entropies at higher O/U ratios. At these ratios, values of $-\Delta \bar{S}_{O_2}$ would be expected to go to $\sim$ as $x \to 0$, and accordingly, Roberts[24] has assumed the presence of a comparatively large number of uranium vacancies. These vacancies are not at equilibrium but are determined by such circumstances as the method of cooling, annealing temperature and rate of cooling.

Thermodynamic results for the high-temperature single-phase UO_{2-x} region have been obtained in recent years[25] and seem reasonably well-established. A point of interest is the very considerable change in oxygen potential close to stoichiometric composition. In Fig. 18 the $\Delta \bar{G}_{O_2}$ values are plotted versus O/U ratio for 2000 °K; this shows a steep fall in $\Delta \bar{G}_{O_2}$ ($= RT \ln p_{O_2}$) close to the stoichiometric composition. A large difference between the $\Delta \bar{G}_{O_2}$ values for UO_{2-x} and UO_{2+x} is to be expected since oxygen atoms are going into vacant anion sites when O/U < 2.0 and into interstitial positions in the lattice when O/U > 2.0. The consequence is that the UO_{2+x} phase is very difficult to define precisely in the region close to stoichiometry and that uranium dioxide in this region of composition is a powerful reducing agent.

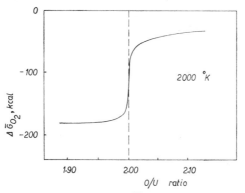

Fig. 18. The partial molar free energy ($\Delta \bar{G}_{O_2}$) as a function of the O/U ratio of UO_{2+x} at 2000 °K.

Indeed, Markin et al.[19] have defined arbitrarily $UO_{2.000}$ as the composition which is in equilibrium with a 10/1 mixture of CO/CO_2 at 850 °C, which has an oxygen potential $\Delta \bar{G}_{O_2} = -100$ kcal/mole. From the partial heats and entropies, the integral values for the reaction:

$$\tfrac{2}{x} UO_{2.00} + O_2 \rightleftarrows \tfrac{2}{x} UO_{2+x}$$

can be calculated at any temperature, from the relationship:

$$\Delta H_T = \frac{1}{x} \int_0^x \Delta \bar{H}_{O_2} \, dx$$

Thus for the reaction:

$$8 \, UO_2 + O_2 \rightarrow 2 \, U_4O_9$$

the value $-\Delta H_{298} = 83.3$ kcal has been found; it is in good agreement with calorimetric measurements of the heat of combustion[18].

The $UO_{2.01} - U_4O_9$ *region* is better known than the single-phase region because the composition does not have to be defined precisely. Average values of $\Delta\bar{G}_{O_2}$[17] for this region are listed in Table 7.

TABLE 7

$\Delta\bar{G}_{O_2}$-VALUES OF THE EQUILIBRIUM UO_{2+x}, U_4O_{9-y}

Temperature, °C	$\Delta\bar{G}_{O_2}$ (kcal)
550°	−55.3
700	−49.5
850	−42.7
950	−38.0
1050	−31.5
1100	−27.4

Low-temperature heat capacity measurements on UO_2 have been made by Jones et al.[26] and later by Westrum et al.[27]; the latter confirmed the thermal anomaly observed by the first mentioned authors and definitely placed it at 30.4 °K. For the entropy of UO_2, it follows that $S^0_{298} = 18.41$ cal/deg.mole.

High-temperature data up to the melting point of UO_2 have been obtained recently, or are being obtained[28,29,30,31]. High-temperature enthalpies have been measured by Hein et al.[28,29], using a drop calorimeter, who found:

$$H_T - H_{298} = -11,688 + 31.937\,T - 9.573 \cdot 10^{-3}\,T^2 + 2.577 \cdot 10^{-6}\,T^3$$

$$(1200–3115\,^0K)$$

A sharp discontinuity in the enthalpy curve at the melting point indicates 18.2 ± 0.5 kcal/mole for the heat of fusion. A somewhat different value (25.3 kcal/mole) has also been found[30].

Preparation and densification of UO_2

Uranium dioxide is generally made by adopting the following major steps: preparation of the starting material (which is usually ADU or uranyl nitrate), calcination of this at about 450 °C to the oxide UO_3, and finally reduction of UO_3 at 650–800 °C to UO_2. Since the UO_2 powder is not thermodynamically stable in contact with oxygen, it is readily oxidized. Indeed, when it is in a finely divided condition, and thus has a high specific surface area (> 10 m²/g), the powder may be either pyrophoric or at least oxidized rapidly at room temperature. In that event, much heat is evolved (25.7 kcal/mole UO_2) and the final product is U_3O_8.

Powders with a lower surface area (< 5 m²/g) are not oxidized to U_3O_8, but to an oxide with an O/U ratio which depends on the surface area; this point will be discussed in more detail in the next paragraph. When employed in nuclear technology, the uranium dioxide must be a high-density material which is close to the stoichiometric composition. This is not only to obtain a high uranium concentration in the reactor core, but also because the physical properties of the oxide, in particular its thermal conductivity, is lower and fission gas release more harmful when densities are lower and O/U ratios higher than the theoretical values.

Conventional techniques, such as cold pressing and sintering, have been successfully applied to obtain UO_2 bodies with a density of more than 95% of that theoretically expected.

Originally, many difficulties had to be met in order to ascertain optimum conditions for the fabrication of close-tolerance, dense UO_2 pellets in large quantities. It appeared that a large number of variables influence the behaviour of UO_2 during sintering. Besides the reduction temperature, variables such as occur in the ball-milling operation, in the O/U ratio, in the compacting pressure and sintering atmosphere, all appear to have an important bearing on the final result. For instance, according to Williams et al.[32], there is a distinct dependence of the sintered density on the O/U ratio in the region O/U = 2.0 – 2.08, whereas a further increase in the O/U ratio does not affect this density. Another interesting observation is the influence which small amounts of gas evolved during the sintering UO_2 may have on the final product[33].

Too much carbon in the green pellets may lead to carbon monoxide formation during sintering and, gives rise to "bubble precipitation" and gross swelling of the compact. This behaviour is also sensitive to deviations in the O/U ratio of the material.

Because of the economical importance of the uranium dioxide fabrication, numerous investigations have been made with results which do not, however, always agree[1]. Obviously there is no simple relationship to describe the behaviour of uranium dioxide powders towards sintering. Moreover, it has become evident that the preparative history of the oxide plays an important role in determining the properties of the final powder, the various steps in the preparation all contribute to its sinterability[34,35,36]. This is so because most of the chemical changes brought about in the solid uranium compounds occur pseudomorphically; thus the size and shape of the UO_2 particles are largely determined by those of the starting material. The morphology and the defects in the structure (texture) in turn determine the reactivity of uranium oxide powders. For this reason, it is not always possible to compare the results of the various authors, especially when they used different starting materials.

To summarize, uranium dioxide powders can be prepared by calcining the starting material (uranyl nitrate or ADU) at about 450 °C, followed by reduction at 650–800 °C. This produces UO_2 powders with a specific surface area of 3–5 m^2/g, and O/U ratios of 2.05–2.12. However, the starting material (see the precipitation process, p. 108) is of great importance since the properties of the UO_2 powder are to a great extent established at this stage. These powders, after cold pressing under a compacting pressure of about 2.8 ton/cm^2, can be sintered to high densities under nitrogen or hydrogen at temperatures of 1400–1700 °C. When nitrogen gas is used a final hydrogen soaking is necessary to ensure the optimum stoichiometry of the final material.

Dense uranium dioxide can also be prepared by electrolysis of a solution of uranyl chloride in a molten salt bath. The reactions involved in the electrodeposition of UO_2 are:

$$UO_2^{2+} + 2\,e \rightarrow UO_2 \quad \text{(cathode)}$$

$$2\,Cl^- \rightarrow Cl_2 + 2\,e \quad \text{(anode)}$$

Various modifications of the process have been described, the salt bath originally[37] was a mixture of NaCl/KCl and operated at a temperature 675°–700 °C; but owing to the decomposition of uranyl chloride other salt mixtures, such as $KCl/PbCl_2$[38] or $KCl/LiCl$[39] which operate at much lower temperatures (450–500 °C) have been employed. The decomposition potential is 0.5 V and in very dry conditions single crystals of UO_2 up to 3 mm across have been made[38]. Only when hydroxide or oxide impurities are present in the electrolyte are branched crystals formed.

For the fabrication of UO_2 in the form of rods for fuel elements, the UO_2 crystals can be densified by vibratory packing or by swaging[1]. The densities obtainable in these ways depend largely on the physical characteristics of the crystals, in particular on the particle size distribution, and densities of even more than 90% can be reached. Melting the UO_2 is probably the best method of producing a material which is suitable for vibrational compaction. Melting can be done in an electric arc furnace; the particles of crushed material are afterwards annealed in hydrogen to obtain O/U ratios close to stoichiometry.

Large single crystals of uranium dioxide (about 1 cm long) can also be obtained by the sublimation of UO_2 at temperatures of 1700 °C and above. The crystals are deposited on the cooler parts of the apparatus and they have O/U ratios which are reported[40] to be nearly stoichiometric. Evidence for the formation of metallic inclusions has not been forthcoming.

Properties of UO_2

The chemical properties of uranium dioxide have been extensively examined, particularly in respect of its oxidation. As already described, the stable oxide in contact with oxygen of 1 atmosphere pressure is UO_3, below 500 °C, and U_3O_8 above. But only very small particles of UO_2 ($\sim$ 100 Å) are oxidized to UO_3 as was demonstrated by infrared absorption analysis[41]. Apparently, the rate of formation of UO_3 is greatly reduced by a kinetic barrier; for instance, finely divided, pyrophoric UO_2 yields U_3O_8.

Some oxidation of uranium dioxide occurs even at very low temperatures: at liquid nitrogen temperatures chemisorption of oxygen has been observed, but up to about 50 °C surface oxidation alone occurs and the diffusion of oxygen ions through the surface layer is the rate controlling factor[41,42,43]. Surface oxidation is responsible for the non-stoichiometric character of UO_2 which has been in contact with dry air at room temperature. Under these circumstances the O/U ratio gradually increases with the time of exposure, and reaches a limiting value of O/U = 2.33. But obviously homogeneous oxidation products cannot be obtained in this way. The UO_2 powders can be stabilized by using higher reduction temperatures or by introducing small amounts of water vapour into the hydrogen. This leads to a protective layer of higher oxide on the UO_2 particles and prevents their further oxidation.

Above 60 °C bulk oxidation begins; this can take place, depending on the surface area, either in one or two stages. Oxides with a surface area of > 1 m^2/g oxidize in two steps. The first step is oxidation to U_3O_7 and is controlled by oxygen diffusion.

The second step which occurs above 200 °C is the oxidation to U_3O_8, and is controlled by a process of nucleation and growth. Differential thermal analysis (DTA) diagrams of UO_2 powders with a very small surface area ($<$ 0.05 m^2/g) showed only one strongly exothermic peak at about 440 °C[44].

Other important chemical aspects of uranium dioxide are its reaction with hydrogen fluoride to give UF_4 (see p. 141) and with other reagents in which it dissolves. The rate of dissolution of the dioxide in HBr, HCl, H_2SO_4 and HNO_3 depends on the particle size and the acid concentration; it is generally low at room temperature. But in warm, strongly oxidizing agents, such as a mixture of HNO_3 and HF, dissolution is rapid.

The physical properties of uranium dioxide are of considerable importance, especially at high temperatures, in relation to its applications as a nuclear fuel. The thermal expansion of the oxide up to its melting point has been measured[45]; the density of solid and liquid UO_2 at the melting point (2800 °C) have the values 9.67 and 8.74 respectively.

The electrical conductivity of nearly stoichiometric single crystals and polycrystalline UO_2 has been measured from room temperature to 3000 °C[2,46]. The electrical conductivity of polycrystalline material appears to be the same as that of single crystal UO_2[46]. Below 1100 °K the conductivity is not reproducible; this is attributed to changes in stoichiometry. As the temperature is raised, the electrical conductivity increases, and rapidly above 1250 °K where the conductivity changes from *p*-type to *n*-type as measurements of the Seebeck coefficient have shown[2]. The electrical conductivity becomes intrinsic above 1400 °C, showing a sharp increase at high temperatures. The results below 1900 °K are best fit by the expression:

$$\sigma = 3.57 \cdot 10^3 \exp(-1.15 \text{ eV}/kT),$$

above 1900 °K by the expression:

$$\sigma = 2.104 \cdot 10^{-2} \cdot T^{1.40} \exp(-0.916 \text{ eV}/kT)$$

It has been suggested[2] that the classical band theory is inadequate to describe the conductivity mechanism. Now, it is assumed that it occurs by the jumping of holes (or electrons) from one ion to a neighbouring ion. The holes are localized at the uranium cations (U^{5+}) and are not free to move through the lattice at low temperatures since they are bound to the interstitial oxygen ions. At high temperatures, however, the holes are free and can move through the lattice. Thus the number of carriers is not increased by raising the temperature, only the mobility of the carriers increases. This is a thermally activated process whereby the positive hole jumps from the U^{5+} to be adjacent U^{4+} cation.

One of the most important physical properties of uranium dioxide is thermal conductivity, since this property is a limiting one in the performance potential of a nuclear-fuel element. Unfortunately, the thermal conductivity of UO_2 is low; it decreases as the temperature is raised, according to a relationship of the type

$$K = A/(B+T),$$

where K is the thermal conductivity and A and B are constants. As might be expected, the influence of its density and deviations from stoichiometry, and of temperature on the thermal conductivity of uranium dioxide have been examined in great detail. The results have been summarized several times[1,2,47]. For instance, the experimental thermal conductivity of sintered, stoichiometric UO_2, in an argon atmosphere, is expressed by the relationship[47]:

$$K = 0.0130 + \frac{1}{T(0.4848 - 0.4465 \cdot D)}$$

$$(800-2000\,°C)$$

in which K is the thermal conductivity (W/cm, $°C$) and D the fraction of theoretical density within the limits 82 and 95%. The equation shows that the thermal conductivity increases rapidly with increasing density (Fig. 19). This is in agreement with what would be expected from the theory, in which the predominant heat-transfer mechanism in uranium dioxide is assumed to be lattice conduction. Such conduction will be reduced by the presence of second phases, local changes in density, and impurities, all of which cause anharmonic disturbances leading to phonon scattering and to a reduction in conductivity. For the same reason, the thermal conductivity falls sharply as the O/U ratio is raised. For the thermal conductivity of polycrystalline $UO_{2.005}$ of 95% of theoretical density, it has been recommended[47a]:

$$K = \frac{1}{11.75 + 0.0235 \cdot t} \quad (W/cm, °C)$$

$$(25-1300\,°C)$$

The thermal conductivity was found to decrease by neutron irradiation, especially below 500 °C.

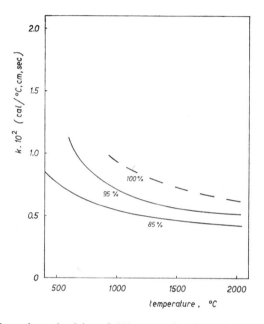

Fig. 19. Thermal conductivity of UO₂ as a function of temperature and percentage of theoretical density (after ref. 47).

Since single crystals of UO₂ sufficiently large to permit accurate conductivity measurements have become available, it has been found that the conductivity of such material is greater than that of ordinary oxide in the temperature range 500–1500 °C. When the thermal conductivity of single crystal UO₂ is plotted versus temperature, it is seen to come down at moderate temperatures, as would be expected from the lattice conduction, but to go up again at higher temperatures possibly owing to the onset of infrared transmission. It has been demonstrated that single crystal UO₂ is transparent over much of the infrared region, actually from 3 to 13 μm; this transparency becomes an important factor in heat transfer above 1000 °C. But whether this effect also reflects differences between specimens (impurity content, porosity, subgrain boundary concentration) has not been resolved.

The U_4O_9 phase

The disordered UO_{2+x} phase changes to a new phase at the composition where $x = 0.25$: namely U_4O_9. The transformation involves long-range ordering of the interstitial oxygen atoms and is characterized by a considerable loss of entropy. Extra, very weak lines appear in the X-ray diagram; these are indicative of a super-lattice, arising from the long-range ordering. Neutron diffraction intensity measurements have been made on U_4O_9 crystals[48], which indicate that the oxygen atoms in the composite cell of U_4O_9 occupy exactly the same kind of positions as do the oxygens in the statistic cell of UO_{2+x}. Both X-ray and neutron-diffraction results have led to the conclusion that U_4O_9 has the space group $I\bar{4}3d$ with a lattice parameter of 21.77 Å, or nearly four times that of UO_2. A model representing the oxygen positions has been proposed by Belbeoch[49]; however, this is not to be reconciled with the neutron-diffraction observations of Willis[48].

The thermal expansion of U_4O_9 shows an interesting anomaly between 20° and 75 °C[50]. In this temperature range there is a small contraction of the lattice parameter, and thereafter it expands uniformly with temperature. Also in the same temperature range, a thermal anomaly in the heat capacity has been observed[51]. The nature of this anomaly has been revealed by Belbeoch et al.[50] who have ascribed it to a phase transformation arising from a weak deformation of the cubic cell to a rhomboedral cell with $a = 5.4438$ Å and $\alpha = 90.078°$. This transition occurs at $65 \pm 10 °C$ and corresponds with the λ-type thermal anomaly in the heat capacity curve.

Evidence for a second phase transition in U_4O_9 has been obtained by Blank and Ronchi[52]; using electron diffraction, they found a reversible order-disorder transformation between 550° and 650 °C. The authors were also able to explain the unusual shape of the UO_{2+x} phase boundary in this way (Fig. 14, p. 65).

It may thus be concluded that U_4O_9 has three different crystallographic forms: α-U_4O_9, which is stable up to about 65°C, β-U_4O_9 which is transformed at about 600°C into the disordered γ-U_4O_9 phase. The homogeneity range of α-U_4O_{9-y} is not well established. At room temperature the composition range is very narrow and the value of y probably does not exceed 0.1. At 1123°C γ-U_4O_9 disproportionates to $UO_{2.244}$ (= the disordered UO_{2+x} phase) and $UO_{2.61}$[6].

Low-temperature heat-capacity measurements on α-U_4O_9 in the range 1.6 to 24°K have been made[53] and values significantly higher than that for UO_2 have been found. The difference is assumed to be of magnetic origin. A revised value for the entropy has been given: $S^0_{298} = 83.53$ cal/deg.mole. Specific heat data of U_4O_{9-y} up to 700°C are not available.

The U_3O_7 phases

Oxidation of uranium dioxide at temperatures below 200°C may result in the formation of metastable tetragonal phases in the composition range $UO_{2.3}$ to $UO_{2.4}$. A compound, empirically U_3O_7 was first identified by Jolibois[54] in 1947 and its existence was confirmed later by several authors[1,2].

In addition, two other tetragonal phases with a composition approximately U_3O_7 have also been claimed, but there is disagreement concerning their precise characterization, in particular on the values of the O/U and the c/a ratios.

The α-U_3O_7 phase is generally obtained when the dioxide is oxidized at temperatures below 135°C; the O/U ratio is about 2.30 and the c/a ratio probably less than 1.0. The phase is transformed into β-U_3O_7 above 180°C, a conversion which is complete at O/U = 2.33 ($c/a = 1.033$ and $c = 5.556$ Å).

At still higher temperatures (> 350°C) a third phase γ-U_3O_7 has been found, with $c/a = 1.017$ at 350°C to 1.010 at 650°C. There is no doubt that γ-U_3O_7 has a lower O/U ratio than β-U_3O_7 since the

$$\beta\text{-}U_4O_9 \rightarrow \gamma\text{-}U_4O_9 + U_3O_8$$

transformation occurs between 360 and 460 °C. The γ-U_3O_7 phase in turn decomposes above 550 °C thus:

$$\gamma\text{-}U_3O_7 \rightarrow U_4O_9 + U_3O_{8-z}$$

Heat capacities of α and β-U_3O_7 have been measured by Westrum and Grønvold[55] and a small λ-type thermal anomaly at 30.5 °K has been observed for α-U_3O_7. The heat of formation of β-U_3O_7 was measured calorimetrically by Fitzgibbon et al.[56], who found the value $-\Delta H_{298} = 818.4 \pm 2.8$ kcal/mole. The evaluation of the free energies of formation of the U_3O_7 phases by Westrum[57] shows them to be metastable with respect to U_4O_9 and U_3O_8.

The oxide U_3O_8

The oxide U_3O_8 exists in at least three crystallographic modifications. The phase commonly encountered is α-U_3O_8 which, in 1 atm of oxygen, is the stable uranium oxide above 500 °C. At room temperature it has a c-face-centered orthorhombic unit cell with lattice parameters $a = 6.72$ Å, $b = 11.96$ Å and $c = 4.15$ Å; the space group is C2mm[58]. When heated, the unit cell of α-U_3O_8 gradually approaches a hexagonal form and, in connection with this, phase transition temperatures varying from 150° to 400 °C have been noted[59,60,61]. Girdhar and Westrum[62] have found recently that at 208.5 °C a λ-type transition point occurs which is accompanied by a heat effect of only 41.7 cal/mole. It is thus clear that the transition from orthorhombic to hexagonal U_3O_8 is a second-order phase change, in full agreement with the crystal structure of α-U_3O_8 which is even at room temperature almost indistinguishable from hexagonal, space group P$\bar{6}$2m. Neutron diffraction data show that such a structure is actually obtained above the transition point[63].

A second form of U_3O_8, designated β-U_3O_8, and first described by Hoekstra et al.[64], has an orthorhombic cell with lattice constants $a = 7.05$ Å, $b = 11.42$ Å and $c = 8.29$ Å, and is thus closely related in symmetry to α-U_3O_8. It often occurs as a contamination in α-U_3O_8. Evidence for still another orthorhombic U_3O_8 phase has been found by Karkhanavalla and George[65], who designated this phase δ-U_3O_8, but later Loopstra[66] showed the phase to be identical with β-U_3O_8.

The β-U_3O_8 modification can be prepared by heating the α-phase in oxygen of 1 atm at temperatures of about 1350 °C and slowly cooling the sample (100° per day) to room temperature. Remarkably, the β-phase is rapidly transformed to α-U_3O_8 when heated at 125 °C; an explanation for this paradoxal behaviour has not yet been given, but it is certainly of kinetic origin.

A third polymorph of U_3O_8 is γ-U_3O_8; a phase that can be made only at very high oxygen pressures ($>$ 16,000 atm) and between 200 and 300 °C[67]. It probably has a hexagonal cell with $a = 8.78$ Å and $c = 9.18$ Å.

Composition and thermodynamic properties of α-U_3O_8

Since α-U_3O_8 is the form of the oxide most frequently weighed in gravimetric uranium analysis, knowledge of its stoichiometry is important, particularly in relation to temperature. From several investigations it has emerged that the composition remains close to that represented by U_3O_8 below a temperature of 800 °C[1,68]. But above 800 °C the compound loses oxygen to give the oxide U_3O_{8-z}, in which the value of z is dependent on the temperature and the oxygen partial pressure[68].

Oxygen pressures of the two phase equilibrium U_4O_9/U_3O_{8-z} have been frequently measured[6,69,70] and the agreement between the results is very good. These can best be fitted by:

$$\log p_{O_2}(\text{atm}) = \frac{-16{,}760}{T} + 8.27$$

The O/U ratio at the upper limit of the two-phase boundary is close to the value 2.61; this is the lower limit of the homogeneity range of U_3O_8. It should be noted that U_3O_{8-z}, when cooled in oxygen, takes up the oxygen very rapidly until the stoichiometric composition has been reached.

References p. 93

Low-temperature heat capacity measurements over the range 5 to 350 °K have been made by Westrum and Grønvold[71]. A small λ-type anomaly has been observed with a maximum at 25.3 °K. At room temperature the standard entropy $S^0_{298} = 67.53$ cal/deg.mole. The heat of formation has been measured by means of combustion calorimetry, and the value $-\Delta H_{298} = 854.1$ kcal/mole has been accepted as the most probable one[17, 56].

The U_3O_8–UO_3 system

At least six crystalline polymorphs of the oxide UO_3, and also an amorphous form of this oxide are known. Phase relationships in this complicated system have formed the intricate subject of much research. The results obtained up to 1960 have been adequately summarized by Hoekstra and Siegel[72]. Since that time, however, our knowledge of this system has been expanded, particularly crystal structures of several of the phases have been investigated.

Preparation and crystallographic properties

α-UO_3, $UO_{2.9}$ *and amorphous* UO_3. The brown-coloured substance α-UO_3 can be prepared by heating hydrated uranium peroxide, containing some nitrate ions, at about 450 °C[73]. When, however, the peroxide which is used, has been thoroughly washed, it decomposes pseudomorphically, first to the amorphous form of UO_3, and later, at temperatures between 450 and 500 °C, it crystallizes to the phase $UO_{2.9}$[74].

Originally it was assumed that α-UO_3 had a hexagonal unit cell[75], but recent work[76] has definitely shown that the generally accepted hexagonal structure cannot be correct. Both X-ray and neutron diffraction results led to the conclusion that α-UO_3 is an imperfectly crystalline form of an orthorhombic modification of the oxide, the appreciable broadening of most lines in the powder pattern being due to twinning which occurs on a scale so small that an average pattern is produced. The lattice parameters of the orthorhombic form are $a = 3.913$ Å, $b = 6.936$ Å and $c = 4.167$ Å, but the presence of a number of very weak lines indicates that the crystallographic unit cell is actually a multiple of this rhombic cell. The structure of the orthorhombic cell is, however, unknown.

The oxide UO$_{2.92}$ should be considered as the untwinned form of the α-UO$_3$ phase[76]; it is the oxide obtained by crystallization of the amorphous UO$_3$, material which is not only amorphous to X-rays but also to electrons. The crystallization of UO$_{2.92}$ has been examined in the electron beam[74].

β-UO$_3$, an orange-coloured powder, can be made by heating either ADU produced from uranyl nitrate and ammonia (p. 108), very slowly to 450 °C in air, or uranyl nitrate very rapidly to the same temperature. Both procedures yield poorly crystalline products. These can, however, be heated further at 500 °C in air, which causes a very slow crystallite growth. After several weeks, an ultimate crystallite size of about 600 Å can be reached[77].

The structure of β-UO$_3$ has been investigated both by X-rays and by neutron diffraction[78]. It has a monoclinic unit cell, with lattice parameters $a = 10.34$ Å, $b = 14.33$ Å, $c = 3.91$ Å and $β = 99.03°$. The unit cell contains 10 formula units and the density is 8.25.

γ-UO$_3$ is a bright-yellow substance that can be prepared by slowly heating uranyl nitrate, first to 200 °C, and after homogenization, to 500 °C. Its structure has been found from X-ray powder diffraction data to be monoclinic, pseudotetragonal[79]. The lattice parameters found are $a = b = 6.89$ Å, $c = 19.94$ Å and $γ = 90.34°$. The space group is I4$_1$/amd and it has a density of 8.01. When heated above 90 °C, it assumes tetragonal symmetry.

δ-UO$_3$ can be prepared only by heating the β-form of UO$_2$(OH)$_2$ ($=β$-UO$_3$ · H$_2$O) in air at 375 °C for more than 24 hours. It is red and it has a cubic structure, being isostructural with ReO$_3$; its lattice parameter $a = 4.15$ Å. Wait[80], who first reported its preparation, has given its composition as UO$_{2.82}$, but material of stoichiometric composition can be obtained when sufficient time is allowed.

ε-UO$_3$, which is brick-red, can be made by passing NO$_2$ over the oxide U$_3$O$_8$ at temperatures of 225° to 300 °C[81]. Its structure has been determined by Kovba et al.[82], who showed it to be triclinic with lattice parameters $a = 4.002$ Å, $b = 3.841$ Å, $c = 4.165$ Å, $\alpha = 98°17'$, $\beta = 90°33'$, and $\gamma = 120°28'$; thus it is closely related in symmetry to U_3O_8. The density of ε-UO_3 is 8.73.

A sixth polymorph of UO_3 has been prepared at the relatively high pressure of 30 kilobars and a temperature of $1100\,°C$[83]. It has an orthorhombic structure with $a = 7.511$ Å, $b = 5.466$ Å and $c = 5.224$ Å; the X-ray density is 8.85. This polymorph will be designated here as η-UO_3.

Evidence for still another polymorphous modification of UO_3 has been obtained by Cornman[84]. He obtained a tan-brown substance by rapidly heating hydrated uranyl nitrate at $425\,°C$ in the presence of 0.6 wt. % of sulfamic acid. The material, designated by the author as ζ-UO_3, could not, however, be prepared free from sulphate and its existence as a true polymorph of UO_3 seems therefore doubtful.

A survey of the relevant properties of the UO_3 phases is given in Table 8.

Thermodynamic properties of the UO_3 polymorphs; the phase diagram

The multiplicity of the forms of UO_3 leads to its having a complex phase diagram, and the relative stabilities of the various phases is a matter of considerable interest. Although much progress has been made, knowledge of the thermodynamic properties of these phases is still very incomplete. The heats of formation of the phases have been determined by measuring their heats of solution in nitric acid[77]; the results are listed in Table 8.

Low-temperature heat capacity measurements on α-, β-, and γ-UO_3 have been made by Westrum[57], and the entropies derived from these are given in Table 8. Moreover, recent high-temperature measurements[85] on these phases are in perfect agreement with the heat-capacity data obtained by Westrum[57] and with the heat-content measurements for γ-UO_3 made by Moore and Kelley[86].

TABLE 8
PROPERTIES OF THE UO_3 POLYMORPHS

	Colour	Structure	Z	Density	$-\Delta H_{298}$ (kcal/mole)	S^o_{298} (cal/deg. mole)
α-UO_3	brown	orthorhombic, $a = 3.913$, $b = 6.936$ Å, $c = 4.167$ Å	—	8.4 (X-ray) 7.2 (exp.)	291.8	23.76
β-UO_3	orange	monoclinic, $a = 10.34$ Å, $b = 14.33$ Å, $c = 3.91$ Å, $\beta = 99.03°$	10	8.25	292.6	23.02
γ-UO_3	yellow	monoclinic, pseudotetragonal, $a = b = 6.89$ Å, $c = 19.94$ Å, $\gamma = 90.34°$	—	8.01	293.5	22.97
δ-UO_3	red	cubic, $a = 4.15$ Å	—	—	290	—
ε-UO_3	brick-red	triclinic, $a = 4.002$ Å, $b = 3.841$ Å, $\alpha = 98°17'$, $\beta = 90°33'$, $\gamma = 120°28'$	—	8.73	291.8	—
η-UO_3	?	orthorhombic, $a = 7.511$ Å, $b = 5.466$ Å, $c = 5.224$ Å	4	8.85	—	—

* References, see text.

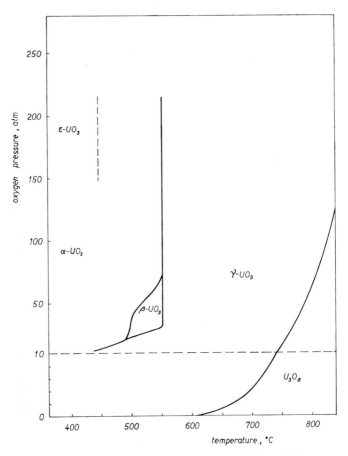

Fig. 20. Phase diagram of the U_3O_8–UO_3 system (after ref. 87).

This information allows the relative stabilities of these phases to be calculated.

An interesting observation is that upon prolonged heating at about 500 °C in air both α-UO_3 and β-UO_3 are converted to γ-UO_3[87]. This indicates that under these circumstances γ-UO_3 is more stable than α-UO_3 and β-UO_3. The dissociation pressure of γ-UO_3, according to the reaction:

$$3 \gamma\text{-}UO_3 \rightleftarrows U_3O_8 + \tfrac{1}{2}O_2$$

has been measured as a function of temperature[87] and the results are:

$$\log p_{O_2}(\text{mm}) = \frac{-12,338}{T} + 16.192$$

From them, the heat of formation of γ-UO_3, $-\Delta H_{298}$ proves to be 293.5 kcal/mole. In agreement with observations by Boullé and Dominé-Bergès[88] the dissociation of UO_3 was found to be reversible. The reoxidation of U_3O_8 is, however, even much slower than the decomposition and leads to what is apparently a form of UO_3 stable at 1 atm of oxygen, namely γ-UO_3.

Some more light has been thrown on the phase diagram by the high-pressure oxidation of U_3O_8[72,89,90]. From these experiments the tentative phase diagram, shown in Fig. 20, has been drawn.

Volatile uranium oxides

The volatile oxide of uranium, observed under oxidizing conditions, is the oxide UO_3. Its thermodynamic properties have been obtained from a study of the equilibrium:

$$U_3O_8 + \tfrac{1}{2}O_2 \rightleftarrows 3\ UO_3(g)$$

Thus for the free energy of formation of $UO_3(g)$ it has been found[91]:

$$\Delta G_T^0 = -198,500 + 19.0\ T$$

Under inert conditions, the vapour phase in equilibrium with uranium dioxide above 2000 °C contains the volatile oxides UO, UO_2 and UO_3. The relative amounts of these oxides which are formed depend on the temperature and the composition of the solid oxide phase. Mass spectrometric and effusion studies, particularly those by Drowart et al.[92] and by Ackermann et al.[93,94] make it clear that the total vapour pressure at a constant temperature increases as the solid phase is oxidized above the composition $UO_{2.00}$ owing to the rapid rise in the partial pressure of $UO_3(g)$.

Originally, it was assumed[2] that congruent vaporization occurs at the composition $UO_{2.00}$. But Drowart et al.[92] have found good evidence that this takes place at O/U < 2.0, for instance, at O/U = 1.987 for 2250°K. At the congruent composition the total vapour pressure has practically the same value as the UO_2 partial pressure; for this it has been found[92] that:

$$\log p_{UO_2}(\text{atm}) = \frac{-30,850}{T} + 8.60$$

$$(1900\text{--}2400\,°\text{K})$$

Below the composition at which congruent vaporization occurs, the vapour pressure increases again owing to a rapid rise in the partial pressures of UO(g) and U(g) until the lower phase boundary of UO_{2-x} is reached, at which point liquid uranium is formed.

For the UO partial pressure over the two-phase equilibrium $U(l) + UO_{2-x}(s)$ it has been found[92]:

$$\log p_{UO}(\text{atm}) = \frac{-28,280}{T} + 8.19$$

$$(1700\text{--}2100\,°\text{K})$$

and for $U(g)$[96]:

$$\log p_U(\text{atm}) = \frac{-26,210}{T} + 5.920$$

$$(1700\text{--}2340\,°\text{K})$$

There is some disagreement concerning the influence of oxygen on the vapour pressure of uranium, and, consequently, on the heat of sublimation of uranium. Whereas Ackermann et al.[95] have found that small amounts of dissolved oxygen suppress the vapour pressure of uranium (the activity of uranium in $UO_2(s)$ being ~ 0.1), Drowart et al.[96] on the contrary hold that uranium remains at unit activity in the presence of oxygen, the uranium pressures for pure uranium and for U/UO_2 mixtures being equal.

The heat of sublimation of uranium derived by the latter authors, $-\Delta H_{298} = 128.5$ kcal/g.atom, is to be preferred since the value has been confirmed by thermochemical cycles, for instance on the

U–C system[97], and by a recent determination of the uranium activity in oxygen-saturated uranium[98]. Moreover, independent measurements of the uranium vapour pressure by Olson[99] have shown a good agreement with the values by Drowart[92]. It may therefore be concluded that the influence of oxygen on the vapour pressure of uranium, if there is any influence, is very slight.

Binary solid solutions and uranates

Uranium dioxide forms an extensive series of solid solutions with many other metal oxides, in particular with the oxides of zirconium, thorium, niobium and the rare earths. Knowledge of these systems is important since, during the fission of UO_2 or $(U, Pu)O_2$, fission product oxides are formed at concentrations (> 1 wt. %); these are high enough to seriously change the properties of the fuel when they react with it. Adequate reviews of these systems have been written by Belle[1] and by Keller[100] and the facts are summarized below.

Continuous solid solutions have been found to be formed with the oxides ThO_2, CeO_2 and with PuO_2. These systems are of importance since fluorite-type solid solution can also dissolve oxygen at temperatures above $100°C$. A study of the system $(U, Th)O_{2+x}$[101] has shown that it behaves in a similar way to that of UO_{2+x} itself; for instance, the values of the enthalpy, $-\Delta \bar{H}_{O_2}$, show only a very small change with the UO_2 content, indicating that ThO_2 acts as an inert diluent. Furthermore, it has been shown that the lattice dimensions of $(U, Th)O_{2+x}$ solid solutions decrease as the mean uranium valency increases from $+4$ to $+5$, and then increase again as the valency rises from $+5$ to $+5.5$. This, and some magnetic evidence[102], suggest that in UO_{2+x} uranium(V) ions, not uranium(VI), are formed.

The UO_2–PuO_2 system is of great importance in nuclear technology, since it provides the fuel for reactors used for producing energy and also for breeding reactors. The oxides form continuous solid solutions and homogeneous solutions can be prepared by a prolonged heating of the mixed oxides at 1500–1600°C. The melting-point diagram shows the complete solid-solution behaviour Fig. 21[103]. A critical assessment of the thermodynamic properties of the system has been recently made[104].

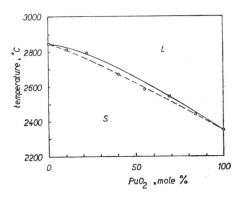

Fig. 21. Phase diagram of the UO_2–PuO_2 system.

With oxides of small cationic radius, for instance, magnesium, beryllium or aluminium, only limited solid solutions have been observed. The oxide MgO can dissolve in the UO_2 lattice up to about 35 mole %, provided that the conditions are sufficiently oxidizing to enable the anionic vacancies to be filled. In the fluorite-type lattice, in which the cationic lattice remains intact, uranium has a formal charge of +5. Similar behaviour has been observed for the UO_2–La_2O_3 system[105] and for the UO_2–Y_2O_3–O_2 system[106]. In the latter system a solid solubility of about 70 mole % $YO_{1.5}$ has been found to give $U_xY_{1-x}O_{1.5+0.5x}$. The solid solution is easily oxidized at room temperature to $(U, Y)O_2$. Apparently the fluorite structure is easily retained as uranium is oxidized to the pentavalent oxidation state. Binary systems of UO_2 with the rare earth metal oxides have been examined by Leitner[107].

The oxide UO_2 is also capable of forming perovskite-type structures with the alkaline-earth metals. For instance, the compound $BaUO_3$ ($a = 4.411$ Å) has been found to exist[108]. It can dissolve up to two moles of BaO per mole of $BaUO_3$ as a solid solution to form $Ba(Ba, U)O_x$, which may have the ultimate composition Ba_3UO_{5+x}. Similar behaviour has been recognized for the strontium compound.

The uranates, in which the uranium is in the hexavalent state, have not been studied systematically. Most uranates have merely been prepared and identified by X-ray examination. They have been generally made by heating the metal oxides, acetates or carbonates with UO_3 or U_3O_8 in air, especially in the large number of preparations made by the Russian authors Spitsyn, Ippolitova, Kovba and coworkers[109]. Their work has considerably extended our knowledge of the uranates, and has provided detailed information on the preparation of many compounds and, in some instances, on their structure and thermal stability.

In the few cases in which all such properties have been systematically examined[110, 111], the complexity of these systems has been revealed. An example is the UO_3–SrO system[110], in which no less than six compounds have been found to exist: SrU_4O_{13}, $Sr_2U_3O_{11}$, α-$SrUO_4$, β-$SrUO_4$, Sr_2UO_5 and Sr_3UO_6. A study of the thermal stability of these phases has shown that their stability increases with increasing Sr/U ratio. At higher temperatures ($> 1000\,^{\circ}C$), or under mildly reducing conditions, the uranates lose oxygen and other structures are produced, which have not been examined in detail.

REFERENCES

1 J. BELLE, Editor, *Uranium Dioxide*, U. S. Government Printing Office, Washington, D.C. (1961).

2 *Thermodynamic and Transport Properties of Uranium Dioxide and Related Phases*, Technical Reports Series, no. 39. I.A.E.A. Vienna (1965).

3 J. M. LEROY AND G. TRIDOT, *Compt. Rend.*, [C] 262 (1966) 114.

3a J. J. KATZ AND E. RABINOWITCH, The Chemistry of uranium. McGraw-Hill Book Company, Inc. New York (1951), p. 246.

4 R. E. RUNDLE, N. C. BAENZIGER, A. S. WILSON AND R. A. McDONALD, *J. Am. Chem. Soc.*, 70 (1948) 99.

5 J. WILLIAMS AND K. H. WESTMACOTT, *Rev. Metallurgy*, 53 (1956) 198.

6 L. E. J. ROBERTS AND A. J. WALTER, *J. Inorg. Nucl. Chem.*, 22 (1961) 213.

7 H. HERING AND P. PERIO, *Bull. Soc. Chim. France*, (1952) 351.

8 F. GRØNVOLD, *J. Inorg. Nucl. Chem.*, 1 (1955) 357.

9 E. F. WESTRUM, Jr. AND F. GRØNVOLD, *J. Phys. Chem. Solids*, 23 (1962) 39.

10   R. K. EDWARDS AND A. E. MARTIN, *Thermodynamics*, Proceedings of a
      Symposium (1965), Vol. 2, IAEA, Vienna (1966), p. 423.

11   P. GUINET, H. VAUGOYEAU AND P. L. BLUM, *Compt. Rend.*, 263 (1966) 17.

12   J. L. BATES, *Thermodynamics*, Proceedings of a Symposium (1965),
      Vol. 2, IAEA, Vienna (1966), p. 73.

13   A. E. MARTIN AND F. C. MRAZEK, *ANL-Report* 7375 (1967).

14   J. L. BATES, *J. Am. Ceram. Soc.*, 49 (1966) 395.

15   B. T. M. WILLIS, *Nature*, 197 (1963) 755.

16   J. G. SCHNIZLEIN, J. D. WOODS, J. D. BINGLE AND R. C. VOGEL, *J.
      Electrochem. Soc.*, 107 (1960) 783.

17   M. H. RAND AND O. KUBASCHEWSKI, *The Thermochemical Properties of
      Uranium Compounds*, Oliver & Boyd, Edinburgh and London (1963).

18   E. J. HUBER, Ch. E. HOLLEY, Jr. AND E. H. MEIERKORD, *J. Am. Chem.
      Soc.*, 74 (1952) 3406.

19   T. L. MARKIN, L. E. J. ROBERTS AND A. WALTER, *Thermodynamics*,
      Proceedings of a Symposium (1962), IAEA, Vienna (1963), p. 693.

20   K. KIUKKOLA, *Acta Chem. Scand.*, 16 (1962) 327.

21   K. HAGEMARK AND M. BROLI, *J. Inorg. Nucl. Chem.*, 28 (1966) 2837.

22   P. GERDANIAN AND M. DODÉ, *Compt. Rend.* [C], 262 (1966) 796.

23   R. J. THORN AND G. H. WINSLOW, *J. Chem. Phys.*, 44 (1966) 2632.

24   L. E. J. ROBERTS AND T. L. MARKIN, *Proc. Brit. Ceram. Soc.*, No. 8
      (1967) 201.

25   T. L. MARKIN, V. J. WHEELER AND R. J. BONES, *J. Inorg. Nucl. Chem.*, 30
      (1968) 807.

26   W. M. JONES, J. GORDON AND E. A. LONG, *J. Chem. Phys.*, 20 (1952) 695.

27   E. F. WESTRUM, Jr. AND J. J. HUNTZICKER, private communication.

28   R. A. HEIN, L. H. SJODAHL AND R. SZWARC, *J. Nucl. Mat.*, 25 (1968) 99.

29   R. A. HEIN, P. N. FLAGELLA AND J. B. CONWAY, *J. Am. Ceram. Soc.*, 51
      (1968) 291.

30   L. N. GROSSMANN AND A. I. KAZNOFF, *J. Am. Ceram. Soc.*, 51 (1968) 59.

31   A. E. OGARD AND J. A. LEARY, *Thermodynamics*, Proceedings of a Sym-
      posium (1967), IAEA, Vienna (1968), p. 651.

32   J. WILLIAMS, E. BARNES, R. SCOTS AND A. HALL, *J. Nucl. Mat.*, 1 (1959)
      28.

33   B. FRANÇOIS, G. KURKA AND R. DELMAS, *Proceedings Second Intern.
      Powder Metall. Conf., Stary Smokovec*, Vol. I (1966), p. 177.

34   H. DOI AND T. ITO, *J. Nucl. Mat.*, 11 (1964) 94.

35   E. H. P. CORDFUNKE AND A. A. VAN DER GIESSEN, *Reactivity of Solids* (5th
      Int. Symp. Munich, 1964), Elsevier Publishing Co., Amsterdam (1965),
      p. 456.

36   E. H. P. CORDFUNKE AND A. A. VAN DER GIESSEN, *J. Nucl. Mat.*, 24 (1967)
      141.

37   R. S. WILKS, *J. Nucl. Mat.*, 7 (1962) 157.

38   F. A. SCOTT AND L. K. MUDGE, *J. Nucl. Mat.*, 9 (1963) 245.

39   M. SCHLECHTER, J. KOOI, R. BILLIAU, R. A. CHARLIER AND G. L. DUMONT,
      J. Nucl. Mat., 15 (1965) 189.

40   W. VAN LIERDE, R. STRUMANE, E. SMETS AND S. AMELINCKX, *J. Nucl.
      Mat.*, 5 (1962) 250.

41 H. HOEKSTRA, A. SANTORO AND S. SIEGEL, *J. Inorg. Nucl. Chem.*, 18 (1961) 166.

42 P. E. BLACKBURN, J. WEISSBART AND E. A. GULBRANSEN, *J. Phys. Chem.*, 62 (1958) 902.

43 L. E. J. ROBERTS, *J. Chem. Soc.*, (1954) 3332; see also: J. S. ANDERSON, L. E. J. ROBERTS AND E. A. HARPER, *J. Chem. Soc.*, (1955) 3946.

44 D. KOLAR, E. D. LYNCH AND J. H. HANDWERK, *J. Am. Ceram. Soc.*, 45 (1962) 141.

45 J. A. CHRISTENSEN, *J. Am. Ceram. Soc.*, 46 (1963) 607.

46 J. L. BATES, C. A. HINMAN AND T. KAWADA, *BNWL-Report* 296, Part I (1966), Part II (1967).

47 R. R. ASAMOTO, F. L. ANSELIN AND A. E. CONTI, *J. Nucl. Mat.* 29 (1969) 67.

47a *Thermal conductivity of uranium dioxide* (Report of a panel), *IAEA, Technical Reports Series*, No. 59, IAEA, Vienna (1966).

48 K. D. ROUSE, T. M. VALENTINE AND B. T. M. WILLIS, *AERE-R-Report* 4414 (1963); see also: B. T. M. WILLIS, *J. Phys. Rad.*, 25 (1964) 431.

49 B. BELBEOCH, C. PIEKARSKI AND P. PERIO, *Acta Cryst.*, 14 (1961) 837.

50 B. BELBEOCH, J. C. BOIVINEAU AND P. PERIO, *J. Phys. Chem. Solids*, 28 (1967) 1267.

51 E. F. WESTRUM, Jr., Y. TAKAHASHI AND F. GRØNVOLD, *J. Phys. Chem.*, 69 (1965) 3192.

52 H. BLANK AND C. RONCHI, *Acta Cryst.*, A24 (1968) 657.

53 H. E. FLOTOW, D. W. OSBORNE AND E. F. WESTRUM, Jr., *J. Chem. Phys.*, 49 (1968) 2438.

54 P. JOLIBOIS, *Compt. Rend.*, 224 (1947) 1395.

55 E. F. WESTRUM, Jr. AND F. GRØNVOLD, *J. Phys. Chem. Solids*, 23 (1962) 39.

56 G. C. FITZGIBBON, D. PAVONE AND Ch. E. HOLLEY, Jr., *J. Chem. Eng. Data*, 12 (1967) 122.

57 E. F. WESTRUM, Jr., *Thermodynamics*, Proceedings of a Symposium (1965), IAEA, Vienna, Vol. 2 (1966), p. 497.

58 B. O. LOOPSTRA, *Acta Cryst.*, 17 (1964) 651.

59 S. SIEGEL, *Acta Cryst.*, 8 (1955) 617.

60 K. J. NOTZ, C. W. HUNTINGTON AND W. BURKHARDT, *Ind. Eng. Chem. Process Dev.*, 1 (1962) 213.

61 W. ERMISCHER, O. HAUSER AND M. SCHENK, *J. Nucl. Mat.*, 16 (1965) 341.

62 H. L. GIRDHAR AND E. F. WESTRUM, Jr., *J. Chem. Eng. Data*, 13 (1968) 531.

63 B. O. LOOPSTRA, *J. Appl. Cryst.*, (1969), to be published.

64 H. R. HOEKSTRA, S. SIEGEL, L. H. FUCHS AND J. J. KATZ, *J. Phys. Chem.*, 59 (1955) 136.

65 M. D. KARKHANAVALLA AND A. M. GEORGE, *J. Nucl. Mat.*, 19 (1966) 267.

66 B. O. LOOPSTRA, *J. Nucl. Mat.*, 29 (1969) 354.

67 W. B. WILSON, *J. Inorg. Nucl. Chem.*, 19 (1961) 212.

68 M. S. RODRIGUEZ DE SASTRE, J. PHILLIPPOT AND C. MOREAU, *CEA-R-Report* 3218 (1967).

69 W. BILTZ AND H. MÜLLER, *Z. Anorg. Allgem. Chem.*, 163 (1927) 257.

70 S. ARONSON AND J. BELLE, *J. Chem. Phys.*, 29 (1958) 151.

71 E. F. WESTRUM, Jr. AND F. GRØNVOLD, *J. Am. Chem. Soc.*, 81 (1959) 1777.

72 H. R. HOEKSTRA AND S. SIEGEL, *J. Inorg. Nucl. Chem.*, 18 (1961) 154.

73 E. H. P. CORDFUNKE, *J. Inorg. Nucl. Chem.*, 23 (1961) 285.

74 E. H. P. CORDFUNKE AND A. A. VAN DER GIESSEN, *J. Inorg. Nucl. Chem.*, 25 (1963) 553.

75 W. H. ZACHARIASEN, *Acta Cryst.*, 1 (1948) 265.

76 B. O. LOOPSTRA AND E. H. P. CORDFUNKE, *Rec. Trav. Chim.*, 85 (1966) 135.

77 E. H. P. CORDFUNKE, *J. Phys. Chem.*, 68 (1964) 3353.

78 P. C. DEBETS, *Acta Cryst.*, 21 (1966) 589.

79 R. ENGMANN AND P. M. DE WOLFF, *Acta Cryst.*, 16 (1963) 993.

80 E. WAIT, *J. Inorg. Nucl. Chem.*, 1 (1955) 309.

81 J. J. KATZ AND D. M. GRUEN, *J. Am. Chem. Soc.*, 71 (1949) 2106.

82 L. M. KOVBA, L. M. VIDAVSKII and E. G. LABUT, *Zhur. Strukt. Khim.*, 4 (1963) 627.

83 S. SIEGEL, H. R. HOEKSTRA AND E. SHERRY, *Acta Cryst.*, 20 (1966) 292.

84 W. R. CORNMAN, Jr., *DP-Report 749* (1962).

85 E. H. P. CORDFUNKE, unpublished results.

86 G. E. MOORE AND K. K. KELLEY, *J. Am. Soc. Chem.*, 69 (1947) 2105.

87 E. H. P. CORDFUNKE AND P. ALING, *Trans. Faraday Soc.*, 61 (1965) 50.

88 A. BOULLÉ AND M. DOMINÉ-BERGÈS, *Compt. Rend.*, 228 (1949) 72.

89 L. M. VIDAVSKII, E. G. LABUT, L. M. KOVBA AND E. A. IPPOLITOVA, *Dokl. Akad. Nauk SSSR*, 154 (1964) 1371.

90 I. SHEFT, S. FRIED AND N. DAVIDSON, *J. Am. Chem. Soc.*, 72 (1950) 2172.

91 R. J. ACKERMANN, R. J. THORN, M. TETENBAUM AND C. ALEXANDER, *J. Phys. Chem.*, 64 (1960) 350.

92 A. PATTORET, J. DROWART AND S. SMOES, *Thermodynamics*, Proceedings of a Symposium (1967), IAEA, Vienna (1968), p. 613.

93 R. J. ACKERMANN, E. G. RAUH AND M. S. CHANDRASEKHARAIAH, *ANL-Report 7048* (1965).

94 R. J. ACKERMANN AND R. J. THORN, *Thermodynamics*, Proceedings of a Symposium, Vienna (1965), Vol. I, IAEA, Vienna (1966), p. 243.

95 R. J. ACKERMANN, E. G. RAUH AND R. J. THORN, *J. Chem. Phys.*, 37 (1962) 2693.

96 J. DROWART, A. PATTORET AND S. SMOES, *J. Chem. Phys.*, 42 (1965) 2629; See also: *Trans. Faraday Soc.*, 65 (1969) 98.

97 E. K. STORMS AND E. J. HUBER, *J. Nucl. Mat.*, 23 (1967) 19.

98 P. E. BLACKBURN AND P. D. HUNT, *ANL-Report 7425* (1967).

99 W. M. OLSON, *Thermodynamics of Nuclear Materials*, Proceedings of a Symposium (1967), IAEA, Vienna (1968), p. 635.

100 C. KELLER, *Über die Festkörperchemie der Actiniden-oxide*, KFK-Report 225 (1964).

101 J. S. ANDERSON, D. N. EDGINTON, L. E. J. ROBERTS AND E. WAIT, *J. Chem. Soc.*, (1954) 3324.

102 M. J. M. LEASK, L. E. J. ROBERTS, A. J. WALTER AND W. P. WOLF, *J. Chem. Soc.*, (1963) 4788.

103 W. L. LYON AND W. E. BAILY, *J. Nucl. Mat.*, 22 (1967) 332.

104 *The Plutonium–Oxygen and Uranium–Plutonium–Oxygen Systems: A Thermochemical Assessment*, Technical Reports Series, No. 79, IAEA, Vienna (1967).

105 D. C. HILL, J. A. HANDWERK AND R. J. BEALS, *ANL-Report* 6711 (1963).

106 K. HAGEMARK, *KR-Report* 48 (1963).

107 L. LEITNER, *KFK-Report* 521 (1967).

108 W. TRZEBIATOWSKI AND A. JABLONSKI, *Nukleonika*, 5 (1960) 587.

109 V. I. SPITSYN, Editor, *Investigations in the Field of Uranium Chemistry*, Moscow (1961), *ANL-Trans*, 33 (1964).

110 E. H. P. CORDFUNKE AND B. O. LOOPSTRA, *J. Inorg. Nucl. Chem.*, 29 (1967) 51.

111 J. G. ALLPRESS, *J. Inorg. Nucl. Chem.*, 26 (1964) 1847.

CHAPTER 7

Uranium Ions and Their Reactions

Introduction

The element uranium has six valence electrons in the configuration $5f^3$, $6d^1$, $7s^2$. The most common oxidation state of $+6$ involves the loss of all these electrons; but the element may also exist in lower oxidation states, and four well-defined oxidation states are known: $+3$ $(5f^3)$, $+4$ $(5f^2)$, $+5$ $(5f^1)$ and $+6$ $(5f^0)$. The radii of these ions are given in Table 9.

TABLE 9

RADII OF URANIUM IONS

U^{3+}	1.03 Å
U^{4+}	0.93 Å
U^{5+}	0.89 Å
U^{6+}	0.83 Å

The magnetic moments observed are primarily due to the 5f electrons on the uranium ions. The sixfold ionized uranium ion is without unpaired electrons and accordingly its compounds should be diamagnetic. This has been confirmed experimentally for compounds like US_2, UN_2 and γ-UO_3 in which only a small intrinsic and temperature-independent paramagnetism has been observed, in accord with the 5f-electron character in the π-bonds of the uranyl ion[1].

Both magnetic and spectroscopic measurements have shown that the U^{5+} ion has the $5f^1$ configuration[2,3,4]. The magnetic properties of the U^{5+} ion in compounds like U_3O_8, Na_3UF_8 and UCl_5 appear to be influenced by the coordination number and the strength of the ligand field[4]. This is particularly so with the U^{4+} ion, for instance in the octahedral crystal field from the six nearest Cl^- or Br^- neighbours[5]. The configuration of the U^{3+} ion has been found to be $5f^3$, in accord with the magnetic susceptibility measurements of, for instance, the uranium trihalides[6].

The chemical properties of uranium are determined to a great extent by the closeness of the energy levels of the 5f and 6d electrons. The 5f radius, large compared with the 4f radius of the corresponding lanthanide ions, causes greater crystal-field splittings than in the lanthanides; it produces an overlap with the electron clouds of close neighbours and a mixing-in of their ligand wave functions, thus giving rise to bonding effects and superexchange[4,7].

The behaviour of the uranium ions in solution is of particular interest in uranium chemistry. For instance, the oxidation/reduction equilibria and the complex forming reactions have not only been a subject of much fundamental research, but have also appeared of great value in a broad field of practical applications. Some examples will be given below to illustrate this.

Oxidation states in aqueous solutions

The oxidation states of uranium in solution have been investigated in detail. Solutions of U^{3+} ions are rose-purple, those of U^{4+} ions are deep-green and those of uranyl ions, UO_2^{2+}, bright-yellow. These solutions have characteristic absorption spectra, which can be used to identify the valence states of the uranium in solution[8]. The oxidation/reduction relationships of the oxidation states are given by the formal potentials (i.e. at equal concentration of uranium in the two oxidation states); this is shown in the potential diagram below[9].

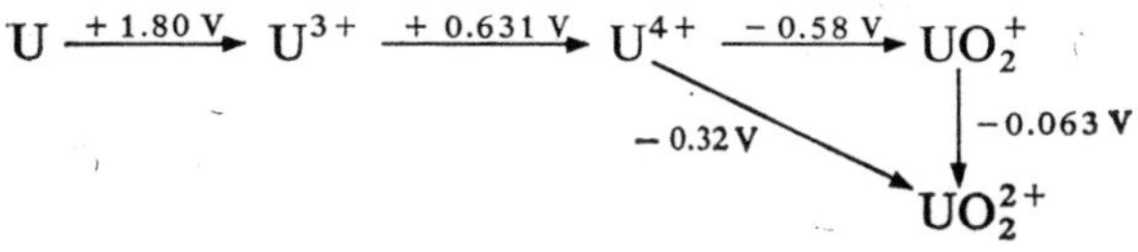

$$U \xrightarrow{+1.80\ V} U^{3+} \xrightarrow{+0.631\ V} U^{4+} \xrightarrow{-0.58\ V} UO_2^+$$

with $-0.32\ V$ and $-0.063\ V$ branches to UO_2^{2+}.

Fig. 22. Formal potentials in 1.0 N $HClO_4$ at 25 °C.

The nature of the uranium ions actually present in the solution must be considered in any interpretation of the potentials. In perchloric acid of moderate acidity hydrolysis is negligible and the ionic species are almost wholly present as hydrated ions. Thus Kritchevsky and Hindman[10] have measured the potential of the U^{4+}/U^{3+} and U^{6+}/U^{5+} couples in perchloric acid by polarographic methods. Both couples appeared to be reversible at the dropping mercury cathode. For the change:

$$U^{3+} \rightarrow U^{4+} + e$$

a formal potential of 0.631 V has been recorded in 1.0 N perchloric acid solution. In 1.0 N hydrochloric acid solution a potential of 0.640 V has been observed for the same couple. The difference between the perchlorate and chloride solution can be attributed to the formation of a weak chloride complex of the U^{4+} ions. Similar observations have been made by Heal[11].

A solution containing U^{3+} ions can be prepared by dissolving a uranium(III) halide in water. The reducing properties of the ions are so strong that hydrogen is evolved from aqueous solutions:

$$2\,U^{3+} + 2\,H_2O \rightarrow 2\,U^{4+} + H_2 + 2\,OH^-$$

As a consequence the properties of U^{3+} ions in aqueous solutions are incompletely known. This is particularly true of the pentavalent UO_2^+ ions which easily disproportionate into a mixture of U^{4+} and UO_2^{2+} ions:

$$2\,UO_2^+ + 4\,H^+ \rightarrow UO_2^{2+} + U^{4+} + 2\,H_2O$$

Thus the equilibrium is dependent on pH. At low uranium concentrations ($\sim 0.001\ M$) and between pH of 2 to 2.5 the rate of disproportionation is reported to be negligibly slow[12, 13].

Ions with pentavalent uranium, UO_2^+, can be obtained, for instance, by photoreduction of uranyl solutions with methanol in perchloric acid. The kinetics of the decomposition of UO_2^+ ions into U^{4+} ions have been measured[14, 15]. The UO_2^+ ions do not form insoluble salts or species extractable with organic solvents and seem to be only slightly hydrolyzed.

Uranium(IV) solutions are generally prepared by the reduction of uranyl salts, for instance, by means of zinc amalgam, or by dissolving uranium(IV) salts, such as UCl_4, in water. The oxidation of aqueous UCl_4-solutions by air oxygen, according to:

$$2\,U^{4+} + 2\,H_2O + O_2 \rightarrow 2\,UO_2^{2+} + 4\,H^+$$

is slow. This is probably because the rate of the oxidation is controlled by the concentration of the hydrolyzed forms of reactants, in particular by the $U(OH)^{3+}$ ion[16]. But the rate of oxidation is appreciably increased by the addition of small amounts of either Cu^{2+} or Fe^{3+} ions[17, 18]. At temperatures between 60° and 80 °C, or in the presence of oxidizing agents, uranium(IV) is easily oxidized. Uranium(IV) ions are extensively hydrolyzed, as is to be expected from the high charge on the ions. Accordingly, very dilute solutions of such ions are not stable when standing in air: they produce a dark precipitate of hydrous UO_2 and the green colour of the solution disappears.

The highest valency of uranium is $+6$. But these ions are, owing to their high charge, unstable in aqueous solutions; they are stabilized by the formation of uranyl ions, UO_2^{2+}. These exist, not only in aqueous solutions, but also in solids, like uranyl nitrate, uranates and hexavalent oxide. Most uranyl compounds are yellow owing to a weak absorption band between 22,000 and 25,000 cm^{-1}.

In the uranyl ion two oxygen atoms are firmly bound to the uranium atom, and the tendency of uranium to form the uranyl group is so great that uranium hexafluoride, for example, is readily transformed in the presence of water into uranyl fluoride, UO_2F_2. Electronic spectra of simple uranyl salts in organic solvents[1] have confirmed the linear structure of the uranyl ion. Indications of this structure had been obtained from spectra in the infrared region[19] and also from the results of X-ray analysis, as will be shown later in this chapter. A discussion of the structure and the spectrum of the uranyl ion and a tentative scheme of the energy levels of the uranyl ion (Fig. 23) has been given by McGlynn and Smith[1]. The complex visible and ultraviolet absorption spectra of the uranyl ion have only recently been resolved[19a].

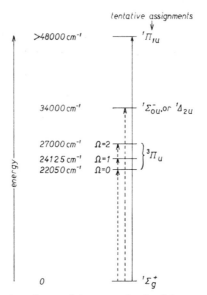

Fig. 23. Tentative scheme of the energy levels of the uranyl ion [1].

The oxidation–reduction relationships of the uranium ions are greatly influenced by complex-formation processes. Thus for the UO_2^{2+}/U^{4+} couple a potential of 0.334 V has been found in perchloric acid solution, whereas in hydrochloric acid solution the potential is 0.407 V. This is because of the greater tendency of the U^{4+} ions than of the UO_2^{2+} ions to form complexes with ligands such as sulphate or chloride. Since the potential of the O_2/H_2O couple is 0.401 V, uranium(IV) ions are stabilized in the presence of complex forming acids, such as sulphuric acid or hydrochloric acid. There are, however, exceptions: the oxidation of uranium(IV) in carbonate solutions, even in a CO_2 atmosphere, takes place very rapidly.

Complex formation

Uranyl ions can form both cationic and anionic complexes. With many oxygen-containing anions, such as carbonate, oxalate and acetate, stable complexes are formed:

$$UO_2(CO_3)_3^{4-} \qquad UO_2(C_2O_4)_3^{4-} \qquad UO_2(C_2H_3O_2)_3^-$$

The coordination number is generally six, and two oxygen atoms of the uranyl group are collinear. These two atoms are closer to the uranium atom (1.7–2 Å) than the six other ligands (2.2–3 Å) and are inert to chemical reaction. Examples are $UO_2(H_2O)_6^{2+}$ and $UO_2Cl_4(H_2O)_2^{2-}$.

In the complex ion $UO_2(CO_3)_3^{4-}$, the carbonate group occupies two coordination positions, as is also the case with, for instance, the oxalato group:

$$[UO_2(C_2O_4)_2(H_2O)_2]^{2-}.$$

The carbonato-complexes have been investigated extensively[20]. The alkali metal salts of the ion $[(UO_2)_2(CO_3)_5(H_2O)]^{6-}$ have all been prepared. The monohydrate of the ammonium salt is a light-yellow substance which is very soluble in water, in which it is hydrolyzed. Thermally the salt is not very stable, in contrast to salts of the tricarbonato complexes $UO_2(CO_3)_3^{4-}$ which are anhydrous and more thermally stable.

The number of ligands coordinated around the uranyl group in all these complexes is six. In many cases, however, steric factors hinder the accommodation of six ligands around the central uranyl ion. For instance, this is so with ligands such as NO_3^- or Cl^-. The formation of penta- and hexa-chloro uranyl complexes encounters great steric hindrance and the compounds have not been prepared, even under the most favourable conditions. In especially dry conditions, the salt $Na_2[UO_2Cl_4]$ has been made, whereas from concentrated hydrochloric acid solutions only the hydrate $K_2[UO_2Cl_4] \cdot 2H_2O$ can be isolated[21]. The ion $UO_2Br_4^{2-}$ is also known to occur in various salts[22]. Ligands, such as NO_3^- and Cl^-, appear to form only very weak complexes with uranyl ions, as is illustrated by the stability constants given in Table 10.

By contrast, the sulphate complexes are much stronger, but the complex $UO_2(SO_4)_3^{4-}$ is unstable in aqueous solutions and can be isolated only with great difficulty as the salt $K_4[UO_2(SO_4)_3] \cdot 2H_2O$. When this salt is dissolved in water, it dissociates rapidly to form the ion $[UO_2(SO_4)_2(H_2O)_2]^{2-}$. This complex is the common sulphato uranyl ion found in aqueous solutions.

TABLE 10

STABILITY CONSTANTS FOR THE COMPLEXES FORMED BY THE
REACTION $UO_2^{2+} + L \leftrightarrows UO_2L$; TEMP. 25 °C; $\mu = 2.0$[23]

UO_2F^+	26
UO_2Cl^+	0.88
$UO_2NO_3{}^+$	0.24
UO_2SO_4	76
$UO_2(SO_4)_2^{2-}$	710

The same instability with respect to water holds for the nitrato complexes. Thus, although the $UO_2(NO_3)_3^-$ ion exists in anhydrous acetone, it cannot be detected in that medium when as little as 10 % water has been added. Its existence in pure nitromethane has been shown by conductometric and spectrophotometric titrations[24].

With neutral, oxygen-containing ligands, such as water, ether, alcohol and ketones, numerous complex compounds have been prepared. It appears that the more polar the group the oxygen atom of which coordinates with the organic molecule, the more stable is the compound formed with uranyl nitrate. The polar group $P=O$ is involved in the formation of the compound of tributyl phosphate with uranyl nitrate. From uranyl nitrate/TBP solutions, the anhydrous solid $UO_2(NO_3)_2 \cdot 2\,TBP$ (m.pt. $= -6\,°C$) has been isolated.

The coordination complexes of that versatile chelating-ligand, acetyl acetone, are very well established; they have the general formula $UO_2(AA)_2 \cdot CH_3OH$[25].

Uranium(IV) ions form complexes analogous to those of the thorium ion. Since the uranium(IV) ion has a high charge and a relatively small radius (0.93 Å), it not only readily forms complexes with various ligands but undergoes hydrolysis as well. Consequently a great diversity of uranium(IV) complexes can exist and the systematic syntheses of a large number of these have been carried out during recent years[20].

As in the uranyl ion, water shows a tendency to direct coordination around the U^{4+} ion. Kraus and Nelson[26, 27] showed that the U^{4+} ions in perchloric acid solutions are hydrated with 6 to 8 molecules of water. Indeed, a coordination number of eight was first found[28] in the tetraoxalato complex $K_4[U(C_2O_4)_4] \cdot 5H_2O$, of which two pairs of optical isomers were isolated later[29].

The complex $U(SO_4)_4^{4-}$ and $U(SO_4)_3^{2-}$ are formed only in solutions with a high concentration of suphate ions. In dilute solutions of the acid, aquo complexes such as $U(SO_4)_2(H_2O)_4$ appear. In addition, the complex ion $U(SO_4)(H_2O)_6^+$ has been found[30, 31]; this indicates that in perchloric acid solutions the ion $U(H_2O)_8^{4+}$ could be expected. With halide ions complex formation is considerably weaker. Thus Ahrland and Larson[32] have found only the monomeric species UCl^{3+} and UBr^{3+}; these have stability constants of 2 and 1.5 respectively.

Hydrolysis

Aqueous solutions of uranium salts have an acid reaction as a result of hydrolysis; for instance:

$$U^{4+} + H_2O \rightleftarrows U(OH)^{3+} + H^+$$

The order of increasing hydrolysis depends on the charge and size of the ion and is as indicated:

$$U^{4+} > UO_2^{2+} > U^{3+}$$

Thus the U^{4+} ion is easily hydrolysed; it also forms much stronger complexes with a given ligand than does the uranyl ion.

The hydrolysis of the uranyl ion has been extensively studied by many investigators, who have proposed a variety of formulae for the species present in acid uranyl salt solutions. According to early cryoscopic measurements by Sutton[33], hydrolysis of the uranyl ion leads to the formation of polymeric ions, such as $U_2O_5^{2+}$, $U_3O_8^{2+}$ and $U_3O_8(OH)^+$. In dilute solutions the existence of the ions $UO_2(OH)^+$ and $(UO_2)_2(OH)^{3+}$ has been suggested[34, 35]. For later stages of the hydrolysis even trimeric species such as $(UO_2)_3(OH)_4^{2+}$, $(UO_2)_3(OH)_5^+$, ... $(UO_2)_3(OH)_8^{2-}$ have been described[33, 35, 36, 37].

In an attempt to account for the various stages of the hydrolysis and the species thus formed, Sillén has postulated a "core-links" mechanism[38]. According to this the uranyl ion builds up sheet-like complexes with double OH-bridges, which may be described by the general formula:

$$UO_2[UO_2(OH)_2]_n^{2+}$$

In these complexes values of n ranging from one to six have been suggested[39]. However, the two main species which have been found are $(UO_2)_2(OH)_2^{2+}$ and $(UO_2)_3(OH)_5^{+}$ [40], a result not expected from the hypothesis of the core-links mechanism since that assumes each link to be added with equal ease and in a hexagonal coordination.

The hydrolytic behaviour of uranyl solutions have been mainly examined by observing potentiometrically the inflections in the pH-curve when OH^- ions are added to a solution containing uranyl ions. But apparently a generally accepted opinion regarding the number and possible formula of the species in solution has not emerged from these studies. It should be pointed out that hydrolytic-polymeric reactions of the uranyl ion have been found to be extremely slow[41] and this renders the attainment of true equilibrium states difficult, which may go some way towards explaining the contradictory results obtained hitherto.

Optical absorption measurements of hydrolysed uranyl salt solutions in chloride and perchlorate media[42] have shown that the molar absorptivities increase sharply with increasing hydroxyl number. These results have given indications of the complexes $(UO_2)_2(OH)_2^{2+}$ and $(UO_2)_3(OH)_5^{+}$ but Baran[43], who also examined the absorption spectra of uranyl solutions, has clearly shown that these spectra remain unchanged within the limits of accuracy as far as it concerns the positions of the various peaks. This implies that the different species—if present at all—cannot be distinguished spectrophotometrically. From this he further concludes that only one uranyl ion species is present in the solutions —a dimeric one he assumes from his observations—and that the only change taking place during the progressive stages of the hydrolysis is that the number of the OH-ligands alters.

A dimeric uranyl ion species has also been postulated by Woodhead et al.[44] who, however, consider it to be an intermediate in the formation of hexameric cations which finally slowly disproportionate to give insoluble uranyl hydroxide, $UO_2(OH)_2 \cdot H_2O$. Woodhead et al.[44] have indeed been able to isolate the nitrate of a hexameric cation as a bright-yellow salt with the formula:

$$[(UO_2)_6(OH)_{12}(H_2O)_{12}H_2] \cdot (NO_3)_2 \cdot xH_2O$$

The salt is stabilized by the water molecules and decomposes when it is heated at 70°C. The authors state that the hexameric uranyl ion, derived from this salt, is identical with the cations previously formulated as $U_3O_8(OH)^+$ by Sutton[33] and as $(UO_2)_3(OH)_5^+$ by Baes et al.[35]. From X-ray diffraction analysis, it has been found that the triclinic unit cell contains six uraniums atoms, and that the cation in aqueous solution has to be considered as a hexamer (rather than a trimer) with a charge of $+2$. The cation contains a closed hexamer of $UO_2(OH)_2$ units, in agreement with a coordination of six around the uranyl ion. Before complete precipitation, colloidal material is formed during the reaction:

$$[(UO_2)_6(OH)_{12}(H_2O)_{12}H_2]^{2+} \rightarrow$$
$$\rightarrow 2\,H^+ + 6\,UO_2(OH)_2 \cdot H_2O + 6\,H_2O$$

Since this reaction is extremely slow, the pH of the solution tends to decrease after a certain quantity of OH^- ions has been added to the uranyl solution.

Hydrolysis of uranium(IV) solutions begins in perchloric acid solutions at an acidity below 1 N. In this medium up to pH = 2, only monomeric $U(OH)^{3+}$ ions are present. But at pH > 2 polymerization has been observed[45, 46]. From this, Sillén and Hietanen[47] assumed the presence of polynuclear species, just as in the case of the uranyl ion, of the general formula:

$$U(UOOH)_n^{n+4}$$

Some typical reactions of uranyl ions

With ammonia

An interesting example of the hydrolytic phenomena, discussed above, is the reaction of ammonia with an aqueous solution of uranyl nitrate. The reaction has been examined by several investigators[41,47,48,49] largely because the final precipitates—the ammonium uranates—are of technological interest, being a starting material in the preparation of technologically important uranium compounds, such as UO_2 or UF_6. The precipitate formed during the reaction is generally referred to as ADU, an abbreviation of ammonium diuranate since the material was formerly supposed to have the formula $(NH_4)_2U_2O_7$. But doubts on this assumption were thrown by that fact that the NH_4/U ratio in the precipitates is usually about 0.5 instead of the theoretical value of 1.0[48]. The complex nature of the so called ADU has been revealed by a systematic study of the ternary system $UO_3-NH_3-H_2O$. In this system four compounds have been found to exist[49], namely $UO_3 \cdot 2H_2O$ (I), $3UO_3 \cdot NH_3 \cdot 5H_2O$ (II), $2UO_3 \cdot NH_3 \cdot 3H_2O$ (III) and $3UO_3 \cdot 2NH_3 \cdot 4H_2O$ (IV). An X-ray analysis of these phases[50] showed the presence of essentially identical hexagonal or pseudo-hexagonal subcells (Fig. 24); the structure itself consists of layers of the composition $UO_2(O_2)$ with the additional oxygen and nitrogen atoms between these layers. The four compounds thus belong crystallographically to the same series and can be represented by the general formula $UO_3 \cdot xNH_3 \cdot (2-x)H_2O$. The compounds I and II are stable in contact with water whereas compounds III and IV are not and can be prepared only in concentrated ammonia.

Knowledge of the ternary system $UO_3-NH_3-H_2O$ has given much better insight into the conditions in which ADU is formed and the effect of these conditions on the product.

When a solution of uranyl nitrate is titrated with ammonia, the precipitates formed in the initial stages of the precipitation, are not in equilibrium with the solution. The pH of the solution decreases with time as a result of a slow hydrolysis of the uranyl ion. When equilibrium is reached the precipitates formed are at pH = 3.5; they consist of $UO_2(OH)_2 \cdot H_2O$ [= ADU(I) or $UO_3 \cdot 2H_2O$]; this material reacts further at a pH between 4 and 7

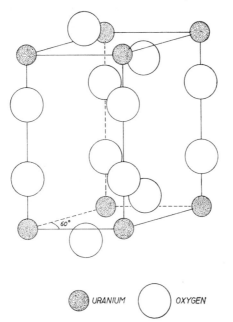

Fig. 24. Idealized subcell of $UO_3 \cdot 2H_2O$[50].

to give the compound ADU-II. At higher pH values a mixture of the compounds II and III is gradually formed. Precipitates formed at a low pH (<7) are generally voluminous and difficult to filter. Moreover they contain variable amounts of nitrate ion absorbed from the solution, and weakly bound as in basic uranyl nitrate[44]. These ions cannot be washed out without causing serious alteration in the composition of the precipitates.

With hydrogen peroxide

When hydrogen peroxide is added to a solution of uranyl nitrate a pale-yellow precipitate with the composition $UO_4 \cdot 4H_2O$ is formed:

$$UO_2^{2+} + 2H_2O_2 + 4H_2O \rightarrow UO_4 \cdot 4H_2O + 4H^+$$

The precipitate, hydrated uranium peroxide, was formerly of some technological interest as a starting material in the production of uranium compounds. The precipitation process is strongly influenced by the concentration of the solution and by its acidity. In solutions with a high concentration of nitric acid, precipitation is very slow owing to the formation of complexes of UO_2^{2+} with peroxide ions, and, as a result, a very crystalline precipitate consisting of large needles may be formed. In dilute, neutral solutions rapid precipitation gives rise to a very voluminous material which is difficult to filter off.

Hydrated uranium peroxide was prepared for the first time by Fairley[51] in 1877. Until recently, however, it was considered to be a peruranate. For instance, Tridot[52] suggested a systematic classification of the perunates as derivatives of the ion of peruranic acid, H_4UO_8, in which hydrated uranium peroxide was formulated as $(UO_2)_2UO_8 \cdot aq$. It was Chernyaev[53] who first described these compounds in terms of the modern concepts of the coordination theory.

Following this lead the bidentate ion OO^{2-} is considered as a ligand, one able to form complexes with the uranyl ion of a stability comparable with that in complexes with CO_3^{2-} ion. Many such complexes have now been isolated[20], for instance, the complexes $(UO_2)_2(OO)_3(H_2O)_6^{2-}$ and $(UO_2)_2(OO)(CO_3)_2^{2-}$, and their properties confirm the analogy just brought out.

Gordon and Taube[54] by tracer studies using ^{18}O, have shown that hydrated uranium peroxide is a true peroxide. Sato[55] has definitely proved the existence of two hydrates, $UO_4 \cdot 4H_2O$ and $UO_4 \cdot 2H_2O$, the former being converted to the latter upon drying in air at temperatures between 80° and 100°C. The equilibrium water vapour pressure between these hydrates has been measured[56] as a function of temperature. The results are expressed thus:

$$\log p_{H_2O}(mm) = \frac{-3056}{T} + 10.936$$

X-ray diffraction evidence on these hydrates has been collected by

Debets[57]. $UO_4 \cdot 4H_2O$ was found to be monoclinic, pseudo-hexagonal, with lattice parameters $a = 11.86$ Å, $b = 6.78$ Å, $c = 4.245$ Å and $\beta = 93° 28'$. For $UO_4 \cdot 2H_2O$ an orthorhombic unit cell was found with $a = 6.50$ Å, $b = 4.211$ Å and $c = 8.78$ Å.

An examination of the thermal decomposition of the hydrates[58] indicates that the tetrahydrate decomposes into the dihydrate in air at about 40 °C and that the dihydrate, in turn, decomposes at about 225 °C to UO_3, O_2 and H_2O.

Uranyl hydroxides–UO_3 hydrates–uranates

Hydrolysis of the uranyl ion yields a yellow substance which is formulated as the hydrate, $UO_3 \cdot 2H_2O$, or as uranyl hydroxide hydrate, $UO_2(OH)_2 \cdot H_2O$. Indeed, addition of ammonia to a solution of uranyl nitrate gives a precipitate which is, at the beginning of the precipitation, essentially $UO_2(OH)_2 \cdot H_2O$. It is the same material as that obtained when UO_3 reacts with water at room temperature, or when UO_2^{2+} ions are hydrolysed by the addition of hydroxyl ions from the anion exchange resin Dowex-2 to a uranyl salt solution. Originally it was designated $UO_3 \cdot 2H_2O$[59]; but, with a growing knowledge of its properties has been variously described as hydrated uranic acid, $H_2UO_4 \cdot H_2O$[60], as hydrated uranyl hydroxide, $UO_2(OH)_2 \cdot H_2O$[61,62] and even as the dimeric structure $U_2O_5(OH)_2 \cdot 3H_2O$[63,64].

Fodor et al.[65], who studied the thermal decomposition of the material, even found indications suggesting a trimeric structure, $[U_3O_7(OH)_4 \cdot 2H_2O] \cdot 2H_2O$.

The "dihydrate" has been assumed to have an amphoteric character, in acid solutions giving ions like UO_2^{2+}, $UO_2(OH)^+$, $U_2O_5^{2+}$ etc, and in alkaline solutions, uranate ions such as UO_4^{2-} and $U_2O_7^{2-}$, for instance, according to the simplified scheme:

$$UO_2^{2+} + 2OH^- \rightleftarrows UO_2\begin{matrix} \diagup OH \\ \diagdown OH \end{matrix} \rightleftarrows UO_4^{2-} + 2H^+$$

Some evidence for a mechanism of this character seemed to be disclosed by the isolation of compounds, such as urea uranate, $H_2UO_4 \cdot 2CO(NH_2)_2$, by Gentile et al.[60]; but it is evident that chemical reactions of this type and the products isolated from them are not very suited to the task of deciding either the kind of ions or the sort of bonding present.

A better insight has been obtained from detailed structural examination. Thus the presence of two OH-groups which are attached to each uranium atom has been shown from infrared absorption measurements[61, 62]. The infrared spectrum of $UO_3 \cdot 2H_2O$ shows, besides the absorption band at 953 cm^{-1} from the asymmetric stretching frequency (v_3) of the uranyl ion and the band at 1610 cm^{-1} from the vibrational bending mode of the coordinated water molecules (δ_{H_2O}), a complex series of bands between 3100 cm^{-1} and 3650 cm^{-1}, due to OH-stretching vibrations. Clearly, $UO_3 \cdot 2H_2O$ is more accurately to be considered as a hydrated uranyl hydroxide. In addition the diffraction pattern of the "dihydrate" has been examined in detail[50, 59, 66]. The substance has an orthorhombic unit cell (space group Pbna) with $a = 13.077$ Å, $b = 16.696$ Å and $c = 14.672$ Å; it has 32 formula units in the cell. The unit cell of this compound (and of the related ADU's) contains almost hexagonal subcells (Fig. 24).

In the UO_3–H_2O system three "monohydrates" have been identified, α-, β-, and ε-$UO_2(OH)_2$. Of these, α-$UO_2(OH)_2$ is the form which is generally encountered. It can be prepared by the dehydration of $UO_2(OH)_2 \cdot H_2O$ by heating it in air at temperatures above 80 °C. In the product the ratio H_2O/UO_3 is always lower than it should be for the composition $UO_2(OH)_2$[59, 65, 67], namely about 0.8, suggesting a composition like $(UO_2)_5(OH)_3O_2 \cdot H_2O$.

The β-modification can be prepared by the hydrolysis of uranyl acetate in water at 110 °C brought about in a sealed glass tube after about 120 hours[68]. In addition, a third form has been found[69]; this has been named ε-$UO_2(OH)_2$ in order to avoid confusion with the older literature[70] in which a γ- and δ- form have also been reported. These reported modifications were shown later to be impure products and not real phases of the system. However, the phase relationships in the system cannot be claimed to be well established.

The compound ε-$UO_2(OH)_2$ when heated in air above 120°C undergoes slow recrystallization into β-$UO_2(OH)_2$. According to Dawson[59] α-$UO_2(OH)_2$ (Dawson's phase I or lath-shaped $UO_3 \cdot H_2O$[71]) is stable in water up to about 180°C; above that temperature β-$UO_2(OH)_2$ (Dawson's phase II or yellow-green $UO_2(OH)_2$[72] or bipyramidal $UO_3 \cdot H_2O$[71]) is formed. This in turn is stable in water to about 280°C. But Harris and Taylor showed that the β-form is easily transformed into the α-form with increase of pressure in the system.

The X-ray powder patterns of the three crystallographically different forms of $UO_2(OH)_2$ have been indexed; the results are summarized in Table 11.

TABLE 11

X-RAY DIFFRACTION DATA ON URANYL HYDROXIDE

	Symmetry	Z	Lattice parameters	X-ray density	Ref.
α-$UO_2(OH)_2$	orthorhombic, space group Pbca	4	$a = 10.19$ Å $b = 6.92$ Å $c = 4.27$ Å	6.71	71
β-$UO_2(OH)_2$	orthorhombic, space group Fmmm	4	$a = 5.635$ Å $b = 6.285$ Å $c = 9.919$ Å	5.71	73
ε-$UO_2(OH)_2$	monoclinic	2	$a = 6.419$ Å $b = 5.518$ Å $c = 5.561$ Å $\beta = 112° 46'$	5.56	69

The preparation of a hemihydrate, $UO_3 \cdot 0.5H_2O$ has also been reported[59], but attempts by later investigators to isolate and characterize it further seem to have been unsuccessful.

Interesting is the existence of a hydrated uranium oxide containing both U(V) and U(VI) ions. The so called "U_3O_8 hydrate" is violet coloured and should be considered as a non-stoichiometric UO_3 hydrate; it has the composition $UO_{2.86} \cdot 1.5H_2O$[59a].

Uranyl ion coordination

Most commonly, uranium is coordinated by six, seven or eight other atoms. In oxygen-containing compounds two diametrically opposite oxygens are situated at closer distances (1.7–2.0 Å) to the central ion than the others (2.2–3 Å), which has led to their designation as "uranyl" oxygens. The other ligands then form a polygon perpendicular to the linear uranyl group. In the case of sevenfold coordination this polygon tends to be a regular, plane pentagon.

The hexagon, occurring for eight-fold coordination is also regular, but puckered.

Seven-fold, pentagonal bipyramidal coordination is relatively rare. It occurs, for instance, in $K_3UO_2F_5$[74], in α-U_3O_8[75], in the high-pressure form of UO_3[76] and, along with octahedral coordination, in β-UO_3[77]. Possibly α-UO_3 is also an example of this coordination[78].

Six-fold, octahedral coordination is more common. It is met in many uranates, for instance, in the alkaline earth uranates, such as BaU_2O_7[80], $BaUO_4$ and β-$SrUO_4$[79, 81] and in β-$UO_2(OH)_2$[73]. Hexagonal-bipyramidal coordination is most common for uranyl salts. It is found, for instance, in $[UO_2(NO_3)_2 \cdot 2H_2O] \cdot 4H_2O$[82], in $CaUO_4$ and α-$SrUO_4$[79], and in $RbUO_2(NO_3)_3$[83]. For a more complete treatment of uranium compounds with seven- and eight-fold coordination the reader is referred to a recent review[84] where many additional references are to be found.

Organometallic uranium compounds; uranocene

Knowledge of the organometallic uranium compounds is very limited, and, until recently, only a few of these compounds have been investigated in any detail. One of them is a red-coloured substance with the approximate composition $(C_5H_5)_3U$ which can be prepared by the reaction of UCl_3 with sodium cyclopentadienide dissolved in tetrahydrofuran, followed by evaporating off the solvent and then heating the dried residue in vacuum[85]. The product, a red sublimate, is obtained in a low yield and is thermally unstable.

A similar reaction, but starting with UCl_4, yielded almost quantitatively the dark-red substance $(C_5H_5)_3UCl$ on heating the residue at 245 °C in vacuum. The properties of the compound indicate that the chlorine atom is ionically bound; they are consistent with its formulation as tri(π-cyclopentadienyl)uranium(IV) chloride, in which three C_5H_5 rings are bound to the metal atom by "sandwich-type" covalent bonds. Spectral data of this compound and of tetracyclopentadienyl uranium have been reported[86].

A quite remarkable example of this type of uranium compounds has been found recently[87]. It is bis(cyclooctatetraenyl)uranium, $(C_8H_8)_2U$, and has a sandwich structure like ferrocene and is called by analogy "uranocene". The unique feature of the material is that the f-orbitals of uranium are shared with the ten π-electrons of the cyclooctatetraene dianion. Evidence for its sandwich structure has been obtained primarily from the mass spectrum. The compound has planar, eight membered rings above and below the central uranium atom in a D_{8d} or D_{8h} arrangement. The substance was prepared by allowing cyclooctatetraene to react with potassium in dry oxygen-free tetrahydrofuran at -30 °C. To the yellow solution at 0 °C, UCl_4 dissolved in tetrahydrofuran is then added. After stirring overnight, followed by the addition of water, green crystals of the compound are obtained in a high yield (80 %). The crystals are sparingly soluble in organic solvents.

The compound $(C_8H_8)_2U$ has an absorption spectrum in the visible range. It inflames on exposure to air, but is unaffected by water, acetic acid and aqueous sodium hydroxide. It is thermally stable and sublimes at 180 °C (0.03 mm).

REFERENCES

1   S. P. McGlynn and J. K. Smith, *J. Molec. Spectroscopy*, 6 (1961) 164.
2   M. J. Reisfeld and G. A. Crosby, *Inorg. Chem.*, 4 (1965) 65.
3   W. Rüdorff and H. Leutner, *Ann.*, 632 (1960) 1.
4   S. Kemmler-Sack, E. Stumpp, W. Rüdorff and H. Erfurth, *Z. Anorg. Allgem. Chem.*, 354 (1967) 287.
5   R. A. Satten, C. L. Schreiber and E. Y. Wong, *J. Chem. Phys.*, 42 (1965) 162.
6   M. Berger and M. J. Sienko, *Inorg. Chem.*, 6 (1967) 324.
7   J. Grunzweig-Genossa, M. Kuznietz and F. Friedman, *Phys. Rev.*, 173 (1968) 562.
8   J. J. Howland in: *Chemistry of Uranium*, collected papers, edited by J. J. Katz and E. Rabinowitz, TID-Report 5290 (1958), Vol. II, p. 680.
9   W. M. Latimer, *Oxidation Potentials*, 2nd Edition, Prentice-Hall Inc., New York (1952), p. 304.
10 E. S. Kritchevsky and J. C. Hindman, *J. Am. Chem. Soc.*, 71 (1949) 2096.
11 H. G. Heal, *Trans. Faraday Soc.*, 24 (1949) 1.
12 F. Nelson and K. A. Kraus, *J. Am. Chem. Soc.*, 71 (1949) 2517.
13 F. Nelson and K. A. Kraus, *J. Am. Chem. Soc.*, 73 (1951) 2157.
14 G. Gordon and H. Taube, *J. Inorg. Nucl. Chem.*, 16 (1961) 268.
15 R. Bressat, B. Claudel, M. Feve and G. Giorgio, *Compt. Rend.* [C], 267 (1968) 707.
16 R. K. Betts, *Can. J. Chem.*, 33 (1955) 1780.
17 A. R. Nichols in: *Chemistry of Uranium*, edited by J. J. Katz and E. Rabinowitz, TID-Report 5290 (1958), Vol. II, p. 450.
18 J. Halpern and J. Smith, *Can. J. Chem.*, 34 (1956) 1419.
19 B. Jezowska-Trzebiatowska, *Nukleonika*, 10, Suppl. (1965) 207.
19a J. T. Bell and R. E. Biggers, *J. Molec. Spectroscopy*, 25 (1968) 312.
20 I. I. Chernyaev, Editor, *Complex Compounds of Uranium*, transl. from Russian, Jerusalem (1966).
21 E. Rimbach, *Ber.*, 37 (1904) 461.
22 J. P. Day and L. M. Venanzi, *J. Chem. Soc.*, (1967) 1363.
23 R. A. Day, Jr. and R. M. Powers, *J. Am. Chem. Soc.*, 76 (1954) 3895.
24 C. C. Addison and N. Hodge, *J. Chem. Soc.*, (1961) 2987.
25 J. P. Fackler, Jr., *Prog. Inorg. Chem.*, 7 (1965) 361.
26 K. A. Kraus and F. Nelson, *J. Am. Chem. Soc.*, 72 (1950) 1391.
27 K. A. Kraus and F. Nelson, *J. Am. Chem. Soc.*, 77 (1955) 3721.
28 P. Pfeiffer, *Z. Anorg. Allgem. Chem.*, 105 (1919) 26.
29 L. E. Marchi and J. Reynolds, *J. Am. Chem. Soc.*, 65 (1943) 333.
30 R. Betts and R. Leigh, *Can. J. Res.*, B28 (1950) 514.
31 R. Day, J. Wilhite and E. Hamilton, *J. Am. Chem. Soc.*, 77 (1955) 3180.
32 S. Ahrland and R. Larson, *Acta Chem. Scand.*, 8 (1954) 137.
33 J. Sutton, *J. Chem. Soc.*, Suppl. Issue, (1949) S 275.
34 S. Ahrland, *Acta Chem. Scand.*, 3 (1949) 374.
35 C. F. Baes, Jr. and N. J. Meyer, *Inorg. Chem.*, 1 (1962) 780.
36 A. Peterson, *Acta Chem. Scand.*, 15 (1961) 101.
37 R. M. Rush, J. S. Johnson and K. A. Kraus, *Inorg. Chem.*, 1 (1962) 378

38 L. G. SILLÉN, *Acta Chem. Scand.*, 16 (1962) 1051.
39 S. AHRLAND, S. HIETANEN AND L. G. SILLÉN, *Acta Chem. Scand.*, 8 (1954) 1907.
40 R. ARNEK AND K. SCHLYTER, *Acta Chem. Scand.*, 22 (1968) 1331.
41 M. E. A. HERMANS, *The Urea Process for* UO_2 *Production*, KEMA, Arnhem (1964).
42 R. M. RUSH AND J. S. JOHNSON, *J. Phys. Chem.*, 67 (1963) 821.
43 V. BARAN, *Z. Chem.*, 5 (1965) 56.
44 J. L. WOODHEAD, A. M. DEANE, A. C. FOX AND J. M. FLETCHER, *J. Inorg. Nucl. Chem.*, 28 (1966) 2175.
45 K. A. KRAUS AND F. NELSON, *J. Am. Chem. Soc.*, 72 (1950) 1391.
46 E. RIMBACH, *Ber.*, 37 (1904) 461.
47 K. J. NOTZ, M. G. MENDEL, C. W. HUNTINGTON AND TH. J. COLLOPY, *TID-Report* 6228 (1960).
48 R. A. EWING, *BMI-Report* 1115 (1956).
49 E. H. P. CORDFUNKE, *J. Inorg. Nucl. Chem.*, 24 (1962) 303.
50 P. C. DEBETS AND B. O. LOOPSTRA, *J. Inorg. Nucl. Chem.*, 25 (1963) 945.
51 T. FAIRLEY, *J. Chem. Soc.*, 31 (1877) 125.
52 G. TRIDOT, *Ann. Univ. Paris*, 24 (1954) 583.
53 I. I. CHERNYAEV, V. A. GOLOVNYA, G. V. ELLERT, R. N. SHCHELOKOV AND V. P. MARKOV, *Proceedings of the Second International Conference on the Peaceful Uses of Atomic Energy, Geneva* (1958), Vol. 28, p. 235.
54 G. GORDON AND H. TAUBE, *J. Inorg. Nucl. Chem.*, 16 (1961) 268.
55 T. SATO, *J. Appl. Chem.*, 13 (1963) 361.
56 E. H. P. CORDFUNKE, *Thermodynamics*, Proceedings of a Symposium, Vienna (1965), IAEA, Vienna, Vol. II (1966), p. 487.
57 P. C. DEBETS, *J. Inorg. Nucl. Chem.*, 25 (1963) 727.
58 E. H. P. CORDFUNKE AND P. ALING, *Rec. Trav. Chim.*, 82 (1963) 257.
59 J. K. DAWSON, E. WAIT, R. ALCOCK AND D. R. CHILTON, *J. Chem. Soc.*, (1956) 3531.
59a E. H. P. CORDFUNKE, G. PRINS AND P. VAN VLAANDEREN, *J. Inorg. Nucl. Chem.*, 30 (1968) 1745.
60 P. S. GENTILE, L. H. TALLEY AND T. J. COLLOPY, *J. Inorg. Nucl. Chem.*, 10 (1959) 114.
61 A. M. DEANE, *J. Inorg. Nucl. Chem.*, 21 (1962) 823.
62 E. SCHWARZMANN AND O. GLEMSER, *Z. Anorg. Allgem. Chem.*, 315 (1962) 305.
63 A. L. PORTE, H. S. GUTOWSKY AND J. E. BOGGS, *J. Chem. Phys.*, 36 (1962) 1695.
64 F. D. LONADIER AND J. E. BOGGS, *J. Less-Common Metals*, 5 (1963) 112.
65 M. FODOR, Z. POKÓ AND J. MINK, *Mikrochimica Acta*, 4/5 (1966) 865.
66 G. KORTLEVE AND P. M. DE WOLFF, *Unpublished results*.
67 H. LANDSPERSKÝ, L. SEDLAKOWA AND D. JAKES, *J. Appl. Chem.*, (1964) 559.
68 G. BERGSTRÖM AND G. LUNDGREN, *Acta Chem. Scand.*, 10 (1956) 673.
69 E. H. P. CORDFUNKE AND P. C. DEBETS, *J. Inorg. Nucl. Chem.*, 26 (1964) 1671.
70 J. J. KATZ AND E. RABINOWITZ, *The Chemistry of Uranium*, MacGraw-Hill Book Company, Inc., New York (1951).

71   L. A. HARRIS AND A. J. TAYLOR, *J. Am. Ceram. Soc.*, 45 (1962) 25.
72   J. PROTAS, *Bull. Soc. Franç. Mineral. Crystallogr.*, 82 (1959) 339.
73   R. B. ROOF, Jr., D. T. CROMER AND A. C. LARSON, *Acta Cryst.*, 17 (1964) 701.
74   W. H. ZACHARIASEN, *Acta Cryst.*, 7 (1954) 794.
75   B. O. LOOPSTRA, *Acta Cryst.*, 17 (1964) 651.
76   S. SIEGEL, H. HOEKSTRA AND E. SHERRY, *Acta Cryst.*, 20 (1966) 292.
77   P. C. DEBETS, *Acta Cryst.*, 21 (1966) 589.
78   B. O. LOOPSTRA AND E. H. P. CORDFUNKE, *Rec. Trav. Chim.*, 85 (1966) 135.
79   E. H. P. CORDFUNKE AND B. O. LOOPSTRA, *J. Inorg. Nucl. Chem.*, 29 (1967) 51.
80   J. G. ALLPRESS, *J. Inorg. Nucl. Chem.*, 27 (1965) 1521.
81   B. O. LOOPSTRA AND H. M. RIETVELD, *Acta Cryst.*, B 25 (1969) 787.
82   J. C. TAYLOR AND M. H. MUELLER, *Acta Cryst.*, 19 (1965) 536.
83   G. A. BARCLAY, T. M. SABINE AND J. C. TAYLOR, *Acta Cryst.*, 19 (1965) 205.
84   E. L. MUETTERTIES AND C. M. WRIGHT, *Quart. Rev.*, 21 (1967) 109.
85   L. T. REYNOLDS AND G. WILKINSON, *J. Inorg. Nucl. Chem.*, 2 (1956) 246.
86   M. L. ANDERSON AND L. R. CRISLER, *J. Organometal. Chem.*, 17 (1969) 345.
87   A. STREITWIESER AND U. MÜLLER-WESTERHOFF, *J. Am. Chem. Soc.*, 90 (1968) 7364.

CHAPTER 8

Uranium Salts

Introduction

Several salts of inorganic acids are known for uranium, in which the element is in the hexavalent oxidation state, either as UO_2^{2+} ions, or tetravalent as U^{4+} ions. Remarkably enough, with a few exceptions, there is little information about these salts; only the composition and the method of preparation are more or less known. In particular, the interesting group of U(IV) salts has been only superficially investigated. Obviously, much further work is needed in this field of uranium chemistry.

Nitrates

Uranyl nitrate, $UO_2(NO_3)_2$, is one of the most important oxosalts in uranium technology, since it is the form in which uranium is purified by the application of solvent extraction to solutions resulting from the processing of the ores. Furthermore, it plays an important role in the reprocessing of fuel elements. For these reasons, its chemistry has been studied extensively.

The phase relationships have been established both by measuring the water-vapour pressures[1] and the solubilities in water[2, 3] of the various phases in the uranyl–water system (Figs. 25 and 26). From this evidence, the existence of the hydrates with 6, 3 and 2 molecules of water as well as of the anhydrous salt has been definitely established. In addition, evidence for the existence of a monohydrate has been obtained.

When uranium or a uranium oxide is dissolved in nitric acid

and the clear solution is evaporated until crystallization begins, the hexahydrate, $UO_2(NO_3)_2 \cdot 6H_2O$, crystallizes upon cooling the liquid to room temperature. The bright-yellow crystals melt

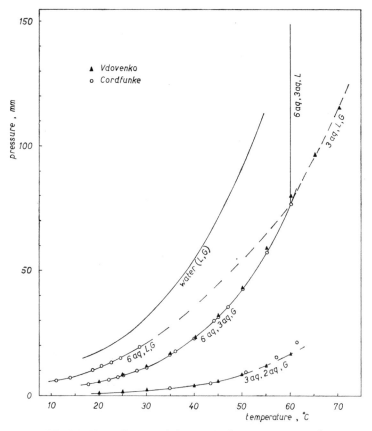

Fig. 25. Phase diagram of the uranyl nitrate–water system[1].

in their crystal water when heated to 60.3 °C[1]; the solid phase in equilibrium with the melt is the trihydrate. The water-vapour pressure of the equilibrium:

$$UO_2(NO_3)_2 \cdot 6H_2O \leftrightarrows UO_2(NO_3)_2 \cdot 3H_2O + 3H_2O_{(g)}$$

has been measured as a function of temperature by Vdovenko et al.[4] and by Cordfunke[1] with results in good agreement.

These may be expressed by the equation:

$$\log p_{H_2O}(mm) = \frac{-2782}{T} + 10.246$$

For the equilibrium:

$$UO_2(NO_3)_2 \cdot 3H_2O \leftrightarrows UO_2(NO_3)_2 \cdot 2H_2O + H_2O(g)$$

it was found:

$$\log p_{H_2O}(mm) = \frac{-3325}{T} + 11.243$$

The phase diagram of the system is shown in Fig. 25, in which is included the water vapour pressure of the saturated solution[1], measured below 30 °C. At higher temperatures the observed pressures are in error because of the evolution of nitrogen oxides

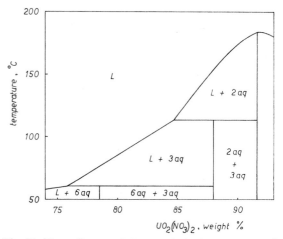

Fig. 26. Phase diagram of the uranyl nitrate–water system[3].

brought about by the hydrolysis of uranyl nitrate. For this reason it is impossible to obtain the pure anhydrous uranyl nitrate by drying the dihydrate at elevated temperatures. Under these circumstances some decomposition occurs as the water is being removed.

The thermal decomposition of uranyl nitrate and its hydrates is of technological importance. The kinetics of the decomposition reaction:

$$UO_2(NO_3)_2 \cdot 6H_2O \rightarrow UO_3 + 2\,NO_2 + \tfrac{1}{2}O_2 + 6H_2O(g)$$

has been studied extensively[5,6]. A mechanism has been proposed[6] in which the decomposition between 200° and 350°C is stated to proceed in two steps, the formation of the monohydrate being the first step. Both reactions are first order and of these the first step is rate-determining. Indeed, evidence for the existence of a monohydrate of uranyl nitrate has recently been obtained by means of X-ray analysis[7].

Uranyl nitrate hexahydrate, $UO_2(NO_3)_2 \cdot 6H_2O$, may be prepared by dissolving uranium or uranium oxide in nitric acid. The solution is evaporated until it begins to crystallize and when cooled to room temperature the hexahydrate separates. It is usually purified by recrystallisation several times from water. When kept over 40% H_2SO_4 the crystals have the correct formula composition.

The hexahydrate has an orthorhombic crystal structure with $a = 13.197$ Å, $b = 8.035$ Å and $c = 11.467$ Å; the space group is Cmc2₁[8]. According to Vdovenko *et al.*[9] it is an assemblage of $UO_2(H_2O)_6^{2+}$ ions, whereas Fleming and Lynton[10] assume that two molecules of water are bonded directly to the uranium atom to give an aquo complex $[UO_2(H_2O)_2(NO_3)_2] \cdot 4H_2O$. This formulation is consistent with the chemical behaviour of anhydrous uranyl nitrate which readily reacts with, for instance, water vapour, NO_2, and organic substances to give compounds with the general formula $UO_2(NO_3)_2 \cdot 2X$. Moreover, it is not possible to dehydrate the dihydrate of uranyl nitrate without decomposition. Obviously, the two molecules of water are directly bonded by proton exchange. This was confirmed in a neutron diffraction study by Taylor and Mueller[8], who found that two molecules of water have a different environment from the four others which are roughly tetrahedrally arranged. The uranyl configuration is shown in Fig. 27; in this the uranyl group is perpendicular to the paper and is surrounded equatorially by an irregular hexagon of six oxygen atoms, four from the NO_3 groups and two from the symmetrically related water oxygens.

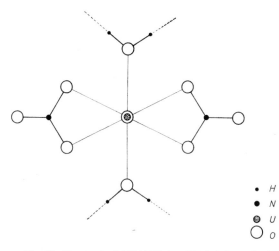

Fig. 27. Structure of $UO_2(NO_3)_2 \cdot 6H_2O$ (after ref. 8).

The heat of formation of uranyl nitrate hexahydrate is found to be $-\Delta H_{298} = 762.3$ kcal/mole[11].

The hexahydrate is a product in most uranium-recovery and purification processes; this is because it is highly soluble in both aqueous solutions and in many organic liquids, such as ethers, alcohols and ketones. The solubility in water has been measured by Marshall et al.[3] and the equilibrium diagram is shown in Fig. 26. For the vapour pressure of the saturated solution in water, it is found[1]:

$$\log p_{H_2O}(mm) = \frac{-2512}{T} + 9.628$$

The high solubility of uranyl nitrate in organic liquids is of practical importance because of its application in the solvent extraction of the salt. It is essential for such an extraction process that the solvent be selective. Of the few nitrates which are extracted from aqueous solutions to an extent comparable with uranyl nitrate, only the latter is extracted under such a wide variety of conditions. Among the liquids which are most selective in the extraction of uranyl nitrate, are the derivatives of phosphoric acid, of which tributyl phosphate (TBP) is used on a large industrial scale.

Because it has a high viscosity and a density near that of water, TBP is diluted with organic solvents when used as an extractant. For this purpose kerosene (30 vol. % TBP in kerosene) is generally employed. Uranyl nitrate is extracted into the organic phase in the form of a neutral, anhydrous complex $UO_2(NO_3)_2 \cdot 2TBP$ (see p. 15).

Uranyl nitrate trihydrate, $UO_2(NO_3)_2 \cdot 3H_2O$, may conveniently be prepared by recrystallisation from a saturated solution of uranyl nitrate, which is $20 N$ with respect to nitric acid. The material maintains the correct water content when kept over 70 % H_2SO_4. The heat of formation of the trihydrate is $-\Delta H_{298} = 550.5$ kcal/mole[11].

Uranyl nitrate dihydrate, $UO_2(NO_3)_2 \cdot 2H_2O$, has a monoclinic structure with lattice parameters $a = 10.52$ Å, $b = 5.93$ Å, $c = 6.95$ Å and $\beta = 72°$[12]. The hydrate is extremely hygroscopic; it may be prepared according to Vdovenko et al.[13] by drying the trihydrate in a vacuum desiccator over concentrated sulphuric acid.

Uranyl nitrate monohydrate, $UO_2(NO_3)_2 \cdot H_2O$, has been isolated recently by Chottard[7] by heating hydrated uranyl nitrate under a water-vapour pressure of 0.1 to 10 mm at temperatures between 100° and 165 °C. The compound has been characterized by X-ray and infrared measurements.

Anhydrous uranyl nitrate cannot be prepared by dehydration of the hydrates, since decomposition takes place along with removal of water. Nevertheless, several investigators have tried to prepare the anhydrous salt[14, 15, 16, 17]. Their results show that it is possible to prepare anhydrous uranyl nitrate by using an intermediate compound $UO_2(NO_3)_2 \cdot 2NO_2$; this is made by the action of nitrogen oxides on the dihydrate. When $UO_2(NO_3)_2 \cdot 2NO_2$ is heated in vacuum at about 175 °C for two hours, the pure anhydrous uranyl nitrate results. When this product is subjected to prolonged heating in vacuum at the same temperature, it gradually loses NO_2, being converted to UO_3.

Anhydrous uranyl nitrate is a light-yellow powder; it is highly reactive, for instance with water and organic substances such as ether. The reaction with liquid ether is vigorous at room temperature. Gibson and Katz[16] observed a weak luminescence in anhydrous uranyl nitrate at 90 °K. However, the coordinate compounds, $UO_2(NO_3)_2 \cdot 2X$, in which X = ether, acetone, dioxane or nitromethane, fluoresce brightly at a temperature of liquid air. A detailed study of the luminescence of these uranyl compounds has been made[18] from which it appears that a necessary condition for the occurrence of luminescence is the coordination of the uranyl ion with an electron donor by which a strong bond is formed.

Sulphates

Uranyl sulphate and its hydrates were first prepared by Klaproth in 1789, early in the development of uranium chemistry. Both the anhydrous salt and the trihydrate were made by Ebelmen[19] in 1842. But, although our knowledge of the uranyl sulphate system has grown since [20], there is still much to be learnt.

The phase relationships in aqueous uranyl sulphate solutions have been examined by Secoy[21, 22] using solubility measurements, and the phase diagram is shown in Fig. 28. Uranyl sulphate solution is stable up to about 300 °C; above this temperature there is separation into two immiscible liquids, a phenomenon rather unusual in inorganic salt solutions. The two phases which are in equilibrium differ much in their uranium concentration and consequently in density, which is the cause of the rapid phase separation.

From this study, it further appears that the solid phases in the system are the trihydrate, the monohydrate and the anhydrous salt. Notz[23], who examined the thermal decomposition of $UO_2SO_4 \cdot 3H_2O$, found that dehydration to the anhydrous salt takes place in three steps. At 120 °C a dihydrate is formed which decomposes at 210 °C to the monohydrate; this in turn is stable to

300 °C and then is converted to the anhydrous salt. Using DTA-analysis, Notz found indications of an enantiotropic phase transition in anhydrous UO_2SO_4 at 755 °C, after which the salt decomposes in a single step to U_3O_8.

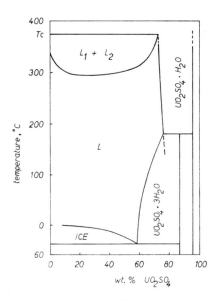

Fig. 28. Phase diagram of the uranyl sulphate–water system (after ref. 21).

It may be noted that the phase relationships in the uranyl sulphate–water system cannot be considered as definitely established. For instance, it has been shown recently[24] that the correct composition of the hydrate, obtained by crystallization from aqueous solutions is $UO_2SO_4 \cdot 2\frac{1}{2}H_2O$. It further appeared that the trihydrate does not exist, whereas the dihydrate found by Colani[25] is formed only as a metastable phase in sulphuric acid solutions of certain concentrations.

When $UO_2SO_4 \cdot 2\frac{1}{2}H_2O$ is heated, it gradually decomposes above 100 °C, passing through an amorphous intermediate, into the anhydrous salt. Indications of the formation of the monohydrate phase did not appear. For the water vapour pressure of the saturated solution of uranyl sulphate in water, it has been found[24]:

$$\log p_{H_2O}(\text{mm}) = \frac{-2229}{T} + 8.783$$

Solid $UO_2SO_4 \cdot 2\frac{1}{2}H_2O$ was without measurable water-vapour pressure below $100\,^{\circ}C$[24]; an observation which does not confirm older measurements by Owens et al.[26].

The ternary system UO_2SO_4–H_2SO_4–H_2O has been investigated and the acid salt $UO_2SO_4 \cdot H_2SO_2 \cdot 2\frac{1}{2}H_2O$ has been identified[24].

$UO_2SO_4 \cdot 2\frac{1}{2}H_2O$ may be prepared by dissolving uranium, or a uranium oxide, in sulphuric acid $(4\,N)$ and evaporating the solution until the beginning of crystallization. After cooling to room temperature crystals of the hydrate separate from the viscous solution; they may be recrystallized from water and dried in air at $50\,^{\circ}C$. As hydrated uranyl sulphate is difficult to crystallize from the very viscous syrups formed on the concentration of aqueous solutions, it is better prepared by placing the anhydrous salt over a saturated solution of sodium chloride. Then the $2\frac{1}{2}$ hydrate is readily obtained with the correct water content.

For the composition $UO_2SO_4 \cdot 3H_2O$ an orthorhombic structure with lattice parameters $a = 12.58$ Å, $b = 17.00$ Å, $c = 6.73$ Å has been reported[26]. However, not only the composition but also the reported X-ray diffraction pattern of this phase seems to be in error[24]. For the heat of formation of $UO_2SO_4 \cdot 2\frac{1}{2}H_2O$, the value $-\Delta H_{298} = 624.4$ kcal/mole has been obtained[27].

$UO_2SO_4 \cdot H_2O$ can be prepared by heating stoichiometric amounts of $UO_2SO_4 \cdot 2\frac{1}{2}H_2O$ and anhydrous uranyl sulphate in a sealed quartz tube at $210\,^{\circ}C$ for 36 hours; the compound is not stable at room temperature[27].

Anhydrous UO_2SO_4 is made by heating any of its hydrates at about $450\,^{\circ}C$. According to Notz[23] it undergoes an enantiotropic transition at $755\,^{\circ}C$, but the X-ray diffraction lines of only the low-temperature (α) form have been reported. From a recent high-temperature X-ray investigation[28], it appears, however, that the phase transition in anhydrous uranyl sulphate is influenced by the presence of small amounts of sulphuric acid as an impurity. In these circumstances, decomposition of the hydrated salt produces only the high-temperature (β) form of UO_2SO_4. The heat of formation of UO_2SO_4 is $-\Delta H_{298} = 441.5$ kcal/mole[27].

U(IV) *sulphate*, $U(SO_4)_2 \cdot 4H_2O$, may be prepared by the reduction of a uranyl sulphate solution. The reaction:

$$UO_2 + 2\,H_2SO_4 \rightarrow U(SO_4)_2 \cdot 4H_2O + 2\,H_2O$$

has been examined in detail by Copenhafer[29], who found that when finely divided UO_2 powder is added to concentrated sulphuric acid at 70–100 °C, in the presence of sufficient water to form the tetrahydrate, and the pasty mixture is maintained at 170 °C, about 75% of the UO_2 powder is converted to $U(SO_4)_2 \cdot 4H_2O$. The extent of the conversion and the time required are primarily dependent on the subdivision of UO_2. It should be noted that the reaction is strongly exothermic and the temperature after mixing rises rapidly.

Uranyl fluoride

Anhydrous uranyl fluoride, UO_2F_2, can be conveniently prepared by the reaction of gaseous hydrogen fluoride on UO_3 at temperatures of 300–500 °C. The reactivity of the type of UO_3 used determines the optimum reaction temperature. The preparation of uranyl fluoride from aqueous solutions is difficult since acid salts are readily formed when UO_3 is dissolved in HF. A study of the system UO_2F_2–HF–H_2O at 25 °C has been reported[30] and, besides the dihydrate, $UO_2F_2 \cdot 2H_2O$, an acid salt of the composition $UO_2F_2 \cdot 2HF \cdot 4H_2O$ has been found. It may be noted that the phase relationships in the uranyl fluoride–water system cannot be considered as definitely established. For instance, besides the well-known dihydrate which exists in three different polymorphs[31], other hydrate phases have been reported. On dehydration at 110 °C, all hydrates lose water to form anhydrous uranyl fluoride; HF is not evolved on dehydration at this temperature.

Uranyl fluoride is readily soluble in water (65.6 wt. % at 25 °C) and in ethyl alcohol but, in contrast to the other uranyl halides, it is not soluble in ether.

Anhydrous UO_2F_2 is a light-yellow substance which is stable in air up to approximately 300 °C; it decomposes completely to U_3O_8 at temperatures of about 800 °C. The compound has a rhombohedral crystal structure (space group R3m), with lattice parameters $a = 5.755$ Å and $\alpha = 42°47'$[32]. The unit cell contains one molecule and the X-ray density is 6.37. For the standard entropy of anhydrous UO_2F_2 it has been found $S_{298}^0 = 32.4$ kcal/mole; the heat of formation $-\Delta H_{298} = 399$ kcal/mole[33].

Uranyl chloride

Uranyl chloride, UO_2Cl_2, was first prepared by Péligot[34] in 1842, who heated UO_2 in a current of dry chlorine. A survey of its properties up to 1945 has been given by Katz and Rabinowitch[35]. Anhydrous uranyl chloride is a bright-yellow substance; it is very hygroscopic and on exposure in moist air forms successively the hydrates $UO_2Cl_2 \cdot H_2O$, $UO_2Cl_2 \cdot 3H_2O$ and, finally, a viscous solution.

The hydrates of uranyl chloride are, in contrast to the anhydrous salt, of a greenish cast and appear to be fluorescent. Their solubility in water is high, but exact values are still lacking. The solubility of the trihydrate in water is reported to be 746 g per 100 g water at 18 °C, and to increase with increasing temperature. For the water vapour pressure of the saturated solution in water, it was found[1]:

$$\log p_{H_2O}(\text{mm}) = \frac{-2100}{T} + 7.80$$

The thermal stability of uranyl chloride and its hydrates has been studied by several investigators. For the water vapour pressure of the equilibrium:

$$UO_2Cl_2 \cdot 3H_2O \rightleftarrows UO_2Cl_2 \cdot H_2O + 2 H_2O(g)$$

it was found[1]:

$$\log p_{H_2O}(mm) = \frac{-3406}{T} + 11.378$$

Prigent and Gueguin[36] have examined the thermal decomposition of $UO_2Cl_2 \cdot H_2O$. They observed that when the hydrate is heated in a current of dry HCl gas, the monohydrate decomposes at 300°C to give the anhydrous salt. When, however, the heating is done in dry nitrogen or oxygen, the basic uranyl chlorides $HU_2O_5Cl_3$ and $U_2O_5Cl_2$ are formed.

According to these authors, anhydrous uranyl chloride is stable up to about 450°C. At higher temperatures it decomposes thus:

$$UO_2Cl_2 \rightarrow UO_2 + Cl_2$$

In order to prevent decomposition from taking place during the preparation, heating in a mixture of hydrogen chloride and chlorine has been suggested[37] for the preparation of anhydrous uranyl chloride.

A knowledge of the thermal stability of UO_2Cl_2 is of interest, since, for instance, the compound is used in a molten salt bath in the electrolytic preparation of UO_2. Its thermal decomposition in a vacuum begins below 400°C, whereas in a HCl/Cl_2 mixture decomposition occurs only above 600°C. In all cases, U_3O_8 is produced in place of UO_2, but its formation has not been explained. Recently, it was found[38] that UO_2Cl_2 reacts with the UO_2 formed through the unstable, intermediate compound UO_2Cl, according to the equilibrium:

$$2 UO_2Cl_2 + 2 UO_2 \rightleftarrows U_3O_8 + UCl_4$$

The formation of pentavalent ions (UO_2^+) in the molten salt mixture had indeed been found polarographically by the same authors.

The compound $UO_2Cl_2 \cdot 3H_2O$ may be made by the slow evaporation of an aqueous uranyl chloride solution. After recrystallization, the product is dried over a saturated solution of lithium chloride. As, according to Mylius and Dietz[39], the basic uranyl chloride, $UO_2(OH)Cl \cdot 2H_2O$, is readily formed in this method of preparation, and as drying over lithium chloride is time-consuming, it is better to prepare the trihydrate by placing the monohydrate of uranyl chloride over a saturated solution of lithium chloride.

$UO_2Cl_2 \cdot 3H_2O$ has an orthorhombic structure with 4 molecules in the cell[40]; its lattice parameters are $a = 12.738$ Å, $b = 10.495$ Å and $c = 5.547$ Å. The space group is Pnma. The heat of formation of the trihydrate has been determined by Shchukarev et al.[41]. With a corrected value for the heat of formation of the anhydrous salt (302.9 kcal/mole[33]), a value for the trihydrate of $-\Delta H_{298} = 521.7$ kcal/mole was found.

The compound $UO_2Cl_2 \cdot H_2O$ is best prepared by treating the residue from an evaporated solution of uranyl chloride with thionyl chloride. The dehydration to give the monohydrate must be done with freshly distilled thionyl chloride; this at the same time converts any basic uranyl chloride which is present into $UO_2Cl_2 \cdot H_2O$. Refluxing with the thionyl chloride speeds up the dehydration[42, 43]. The excess of thionyl chloride is distilled off and the residue is dried in vacuum at 80 °C. The yellow product is the monohydrate in a powder form. It is very hygroscopic and must be handled in a dry-box.

$UO_2Cl_2 \cdot H_2O$ is monoclinic with two molecules in the cell; the lattice parameters are $a = 5.836$ Å, $b = 8.563$ Å, $c = 5.566$ Å and $\beta = 97.70°$; the space group is $P2_1/m$[40]. For the heat of formation, it has been found that $-\Delta H_{298} = 381.8$ kcal/mole[41].

Anhydrous UO_2Cl_2 is conveniently made by dehydrating the monohydrate in a current of a gaseous mixture of HCl/Cl_2 at about 450 °C. Although it is possible to use the trihydrate as the starting material, large amounts of water are then produced with the consequent risk of hydrolysis.

Another method of preparation consists in the oxidation of uranium tetrachloride, UCl_4, in a current of dry oxygen at 300 to 350 °C[44]. The oxidation of UCl_4 may occur in two steps, uranyl chloride being the intermediate product [45, 46]:

$$UCl_4 + O_2 \rightarrow UO_2Cl_2 + Cl_2$$

$$3\, UO_2Cl_2 + O_2 \rightarrow U_3O_8 + 3\, Cl_2$$

References p. 135

It was shown[47] that when a chlorine pressure of about 0.5 atm. is maintained during the oxidation of UCl_4 at 325 °C, the second step does not occur and pure UO_2Cl_2 is obtained. Some UCl_5 is formed by the reaction of chlorine with UCl_4.

The oxidation of UO_2Cl_2 begins at 390 °C; at this temperature the reaction is sluggish and only small amounts of U_3O_8 are formed at the particle surface[48]. At temperatures higher than 440 °C a fast reaction occurs. Bozic and Gal[49] have employed a quite different method, using the reaction of HCl gas with U_3O_8 at 1000 to 1200 °C. It is remarkable that the authors found only slight decomposition at these high preparation temperatures. Uranyl chloride is a bright-yellow substance which melts (in HCl/Cl_2) at 578 ± 3 °C[50]. The salt is extremely hygroscopic and must be handled in a dry box. It has an orthorhombic structure with lattice parameters of $a = 5.724$ Å, $b = 8.408$ Å, $c = 8.719$ Å; the unit cell contains 4 molecules and the space group is Pnma[40].

The heat of formation of UO_2Cl_2, as determined by Shchukarev et al.[51], has been corrected[33] to give a value of $-\Delta H_{298} = 302.9$ kcal/mole. Its entropy, obtained from low-temperature heat-capacity measurements[52], is $S^0_{298} = 36.0$ cal/deg. mole.

The stability of the uranyl halides decreases with increasing atomic number. Whereas $UO_2Br_2 \cdot 3H_2O$ can be prepared by dissolving uranium trioxide in hydrobromic acid[53], uranyl iodide cannot be made in the solid state.

Phosphates

Most investigations on the uranyl phosphates have been concerned with natural products; these are often mixtures, and the results which have been obtained are difficult to interpret. One of the important uranium ores is autunite, a calcium uranyl phosphate with the approximate composition $CaO \cdot 2UO_3 \cdot P_2O_5 \cdot 10H_2O$. A systematic study of the normal, acid and double uranyl phosphates has not been made yet.

Karpov[54] prepared normal uranyl phosphate, $(UO_2)_3(PO_4)_2 \cdot$ $\cdot 6H_2O$, by adding phosphoric acid to a 0.042 M solution of uranyl nitrate in the molar ratio 2:3 at 40°C; he also prepared the acid phosphate, $UO_2HPO_4 \cdot 4H_2O$. The same author studied the thermal decomposition of these phosphates[55] and found that the normal phosphate decomposes through the $4\frac{1}{2}$ hydrate and the monohydrate into the anhydrous uranyl phosphate $(UO_2)_3(PO_4)_2$ at about 275°C. The acid phosphate, however, decomposes through the hemihydrate to uranyl pyrophosphate, $(UO_2)_2P_2O_7$, at 600°C.

Uranium(IV) metaphosphate, $U(PO_3)_4$ can be prepared by the reaction of UO_2 or U_3O_8 with P_2O_5 at temperatures ranging from 400°–1200°C; any excess of P_2O_5 is leached out with boiling water[56, 57]. The compound has two crystallographic modifications; the metastable β-form (orthorhombic) is transformed into the high-temperature α-form (monoclinic) at about 970°C. X-ray data for these polymorphs have been given by Baskin[57]. The α-form begins to decompose into UP_2O_7 at 1200°C.

Carbonates

The solid phases in the uranyl carbonate–water system are difficult to isolate since hydrolyzed products are easily formed. Miller *et al.*[58] prepared for the first time anhydrous uranyl carbonate, UO_2CO_3, by the action of CO_2 under pressure on UO_3 or ammonium uranate. The reaction of CO_2 under pressure with aqueous or alcoholic solutions of uranyl nitrate also yields the anhydrous carbonate[59, 60]. In a quite different method[61] uranium(IV) oxycarbonate, $UOCO_3 \cdot xH_2O$, is first prepared as a green precipitate when ammonium carbonate is added to a photochemically reduced solution of uranyl nitrate. On exposure to dry air this compound undergoes simultaneous oxidation and dehydration with the formation of a yellow hydrate of uranyl carbonate, $UO_2CO_3 \cdot 2\frac{1}{2}H_2O$. When this is heated in a vacuum at 160°C, anhydrous UO_2CO_3 is formed which is stable up to 500°C. A variant of this method has been described by Cejka[62], who first made the violet uranium oxide hydrate, $UO_{2.86} \cdot xH_2O$ (see p. 113), by the photochemical reduction of a uranyl nitrate solution, and then saturated the solution with air and carbon dioxide.

Salts of organic acids

Uranyl formate may be obtained as the neutral salt

$$UO_2(HCOO)_2 \cdot H_2O[63].$$

Its thermal decomposition has been studied recently together with that of uranium(IV) formate, $U(HCOO)_4$,[64].

Uranyl acetate crystallizes from dilute acetic acid solutions as the dihydrate, $UO_2(CH_3COO)_2 \cdot 2H_2O$. The structure is ortho-rhombic with four molecules in the cell. Its lattice parameters are $a = 14.95$ Å, $b = 9.61$ Å and $c = 6.93$ Å[65]. Anhydrous uranyl acetate is formed either by heating the dihydrate at 115 °C or by the action of acetic anhydride on uranyl nitrate[66]; it reacts readily with alcohols to give $UO_2(CH_3COO)_2 \cdot (Alc)_2$. Anhydrous uranyl acetate is stable up to 245 °C at which temperature it begins to decompose slowly[67].

Interesting compounds are the alkali-uranyl acetates of which the sodium salt is the least soluble. The compound, $NaUO_2(CH_3COO)_3$ is anhydrous; it has a cubic structure with $a = 10.688$ Å.

Uranium(IV) acetate is made by photochemical reduction of a solution of uranyl acetate. The salt is anhydrous and has a mono-clinic cell with lattice parameters $a = 17.80$ Å, $b = 8.35$ Å, $c = 8.33$ Å and $\beta = 106° 50'$. The cell contains four molecules and the space group is $C2/c$[68].

Uranyl oxalate, $UO_2(C_2O_4) \cdot 3H_2O$, is obtained by crystalli-zation from aqueous oxalate solutions; its correct composition was first given by Péligot[34]. The salt is only slightly soluble in water, but is dissolved readily in the presence of oxalic acid or ammonium oxalate owing to active complex formation. In the latter case, the double salt $(NH_4)_2(C_2O_4) \cdot UO_2(C_2O_4) \cdot 3H_2O$, is formed[69]. The system $UO_2(C_2O_4)$–oxalic acid–water has been studied by Colani[70], who found the only solid phases between 0° and 70°C to be the trihydrate of uranyl oxalate and oxalic acid dihydrate. $UO_2(C_2O_4) \cdot 3H_2O$ is monoclinic with lattice parameters $a = 5.61$ Å, $b = 17.04$ Å, $c = 9.41$ Å and $\beta = 98° 12'$[71]. On heating it loses two molecules of water at 100°C and the third at 170°C to form the anhydrous salt, which in turn decomposes at about 350°C[72].

An interesting compound is *uranium(IV)oxalate*, $U(C_2O_4)_2 \cdot 6H_2O$. This salt has a green colour; but when contaminated with oxalic acid it is almost white. It is slightly soluble in water and in dilute acids and loses 5 molecules of water when heated at 110°C and the sixth at 200°C[73]. Wendlandt *et al.*[74], who examined the thermal decomposition of this salt, reported the formation of an intermediate dihydrate. The monohydrate decomposes directly to uranium oxide without the formation of the anhydrous salt.

REFERENCES

1 E. H. P. CORDFUNKE, *Thermodynamics*, Proceedings of a Symposium (1965), IAEA, Vienna, Vol. II (1966), p. 483.
2 A. COLANI, *Bull. Soc. Chim.*, 39 (1926) 1243.
3 W. L. MARSHALL, J. S. GILL AND C. H. SECOY, *J. Am. Chem. Soc.*, 73 (1951) 1867.
4 V. M. VDOVENKO AND A. P. SOKOLOV, *Radiokhimiya*, 1 (1959) 117.
5 R. S. ONDREJCIN AND T. P. GARRETT, *J. Phys. Chem.*, 65 (1961) 470.
6 S. HARTLAND AND R. J. NESBITT, *J. Appl. Chem.*, 14 (1964) 406.
7 G. CHOTTARD, *Compt. Rend.*, [C], 267 (1968) 147.
8 J. C. TAYLOR AND M. H. MUELLER, *Acta Cryst.*, 19 (1965) 536.
9 V. M. VDOVENKO, E. V. STROGANOV, A. P. SOKOLOV AND V. N. ZANDIN, *Radiokhimiya*, 2 (1960) 24 (*AEC-tr* 4576).
10 J. E. FLEMING AND H. LYNTON, *Chem. and Ind.*, (1960) 1416.
11 E. H. P. CORDFUNKE, *J. Phys. Chem.*, 68 (1964) 3353.
12 V. M. VDOVENKO, E. V. STROGANOV, A. P. SOKOLOV AND G. LUNGU, *Radiokhimiya*, 4 (1962) 59 (*AEC-tr-982*).
13 V. M. VDOVENKO, E. V. STROGANOV AND A. P. SOKOLOV, *Radiokhimiya*, 3 (1961) 19 (*AEC-tr-4579*).
14 M. MARKETOS, *Bull. Soc. Chim.*, 11 (1912) 144.
15 E. SPÄTH, *Monatsh.*, 33 (1912) 505.
16 G. GIBSON AND J. J. KATZ, *J. Am. Chem. Soc.*, 73 (1951) 5436.
17 V. M. VDOVENKO, I. G. SUGLOBOVA AND D. N. SUGLOBOV, *Radiokhimiya*, 1 (1959) 637.

18   G. I. KOBYSHEV AND D. N. SUGLOBOV, *Dokl. Akad. Nauk. SSSR*, 120 (1958) 330.
19   J. J. EBELMEN, *Ann.*, 43 (1842) 305.
20   *Gmelins Handbuch der Anorganischen Chemie*, 8th edition, *Uran und Isotope*, System Nr. 55 (1936).
21   C. H. SECOY, *J. Am. Chem. Soc.*, 70 (1948) 3450.
22   C. H. SECOY, *J. Am. Chem. Soc.*, 72 (1950) 3343.
23   K. J. NOTZ, Jr., *NLCO-Report* 814 (1960).
24   E. H. P. CORDFUNKE, *J. Inorg. Nucl. Chem.*, 31 (1969) 1327.
25   A. COLANI, *Bull. Soc. Chim.*, 43 (1928) 754.
26   R. J. TRAILL, *Am. Min.*, 37 (1952) 394.
27   E. H. P. CORDFUNKE AND W. OUWELTJES, unpublished results.
28   E. H. P. CORDFUNKE, to be published.
29   D. T. COPENHAFER, *MCW-Report* 10 (1946).
30   YU. A. BUSLAEV, N. S. NIKOLAEV, AND I. V. TANANAEV, *Dokl. Akad. Nauk SSSR*, 148 (1963) 832.
31   W. L. MARSHALL, Jr., J. S. GILL AND C. H. SECOY, *J. Amer. Chem. Soc.*, 76 (1954) 4279; see also: G. NEVEU, *CEA-Report* 2106 (1961).
32   W. H. ZACHARIASEN, *Acta Cryst.*, 1 (1948) 277.
33   M. H. RAND AND O. KUBASCHEWSKI, *The Thermochemical Properties of Uranium Compounds*, Oliver & Boyd, Edinburgh and London (1963), p. 58.
34   E. PÉLIGOT, *Ann.*, 43 (1842) 255.
35   J. J. KATZ AND E. RABINOWITCH, *The Chemistry of Uranium*, Part I, McGraw-Hill Book Company, Inc., New York (1951).
36   J. PRIGENT AND M. GUEGUIN, *Compt. Rend.*, 258 (1964) 4069.
37   D. C. BRADLEY, A. K. CHATTERJEE AND A. K. CHATTERJEE, *J. Inorg. Nucl. Chem.*, 12 (1959) 71.
38   M. V. SMIRNOV, V. E. KOMAROV AND A. P. KORYNSHIN, *Energie Atomique*, 22 (1967) 47.
39   F. MYLIUS AND R. DIETZ, *Ber.*, 34 (1901) 2774.
40   P. C. DEBETS, *Acta Cryst.*, B24 (1968) 400.
41   S. A. SHCHUKAREV, I. V. VASILKOVA, V. M. DROSDOVA AND K. E. FRANTSEVA, *Zhur. Neorg. Khim.*, 4 (1959) 39.
42   J. D. HEFLEY, D. M. MATHEWS AND E. S. AMIS, in: J. KLEINBERG, Ed., *Inorganic Syntheses*, Vol. 7, McGraw-Hill, New-York (1963), p. 146.
43   G. PRINS, private communication.
44   J. A. LEARY AND J. F. SUTTLE, *Inorganic Syntheses*, Vol. 5 (1957), p. 148.
45   J. VAN WAZER AND G. JOHN, *J. Am. Chem. Soc.*, 70 (1948) 1207.
46   B. KANELLAKOPOULOS AND H. PARTHEY, *J. Inorg. Nucl. Chem.*, 28 (1966) 2541.
47   E. F. WESTRUM, Jr., *Thermodynamics of Nuclear Materials*, Proceedings of a Symposium (1962), IAEA, Vienna (1963), p. 19.
48   B. KANELLAKOPOULOS AND H. PARTHEY, *J. Inorg. Nucl. Chem.*, 30 (1968) 1209.
49   B. I. BOZIC AND O. GAL, *Z. Anorg. Allgem. Chem.*, 273 (1953) 84.
50   L. OCHS AND F. STRASSMANN, *Z. Naturforschung*, 7b (1952) 637.
51   S. A. SHCHUKAREV AND YU. G. MALTSEV, *Zhur. Neorg. Khim.*, 3 (1958) 2647.

52 E. GREENBERG AND E. F. WESTRUM, Jr., *J. Am. Chem. Soc.*, 78 (1956) 4526.
53 S. PETERSON, *J. Inorg. Nucl. Chem.*, 17 (1961) 135.
54 V. I. KARPOV, *Zhur. Neorg. Khim.*, 6 (1961) 531; see *Russ. J. Inorg. Chem.*, 6 (1961) 271.
55 V. I. KARPOV AND Ts. L. ANBARTSUMYAN, *Zhur. Neorg. Khim.*, 7 (1962) 1838; see *Russ. J. Inorg. Chem.*, 7 (1962) 949.
56 A. BURDESE AND M. LUCCO BORLERA, *Ann. Chim.*, 53 (1963) 344.
57 Y. BASKIN, *J. Inorg. Nucl. Chem.*, 29 (1967) 383.
58 P. D. MILLER, H. A. PRAY AND H. P. MUNGER, *AECD-Report* 2740 (1949).
59 I. I. CHERNYAEV, V. A. GOLOVNYA AND G. V. ELLERT, *Zhur. Neorg. Khim.*, 1 (1956) 2726.
60 C. FRONDEL AND R. MEYOWITZ, *Am. Min.*, 41 (1956) 127.
61 B. SAHOO AND D. PATNAIK, *Nature*, 185 (1960) 683.
62 J. CEJKA, *Coll. Czechoslov. Chem. Commun.*, 24 (1959) 3180.
63 A. MÜLLER, *Z. Anorg. Allgem. Chem.*, 109 (1920) 235.
64 G. D. BUTTRESS AND M. A. HUGHES, *J. Chem. Soc.* (A), (1968) 1272.
65 V. AMIRTHALINGAM, D. V. CHANDRAN AND V. M. PADMANABHAN, *Acta Cryst.*, 12 (1959) 821.
66 E. SPÄTH, *Monatsh.*, 33 (1912) 235.
67 P. S. CLOUGH, D. DOLLIMORE AND P. GRUNDY, *J. Inorg. Nucl. Chem.*, 31 (1969) 361.
68 I. JELENIC, D. GRDENIC AND A. BEZJAK, *Acta Cryst.*, 17 (1964) 758.
69 I. I. CHERNYAEV, Editor, *Complex Compounds of Uranium*, Moscow (1964); translation Jerusalem (1966).
70 A. COLANI, *Bull. Soc. Chim.*, 37 (1925) 858.
71 E. STARITZKY AND D. T. CROMER, *Anal. Chem.*, 28 (1956) 1353.
72 V. M. PADMANABHAN, S. C. SARAIYA AND A. K. SUNDARAM, *J. Inorg. Nucl. Chem.*, 12 (1960) 356.
73 A. ROSENHEIM AND M. KELMY, *Z. Anorg. Allgem. Chem.*, 206 (1932) 33.
74 W. W. WENDLANDT, T. D. GEORGE AND G. R. HORSTON, *J. Inorg. Nucl. Chem.*, 17 (1961) 273.

CHAPTER 9

Uranium Halides

Introduction

The uranium halides constitute an interesting group of uranium compounds. For most of them, chemical and physical properties have been investigated intensively and the trend in properties both with increasing atomic number of the halogen atom and with increasing valency of the uranium atom has become obvious. For instance, all uranium halides are hygroscopic and sensitive to oxidation by air, but the extent to which they are so, depends on the type of halide. Thus, the sensitivity to air oxidation and to reaction with water, increases from UF_3, which is not very hygroscopic, to UI_3. All the tetrahalides and especially UCl_6 are very hygroscopic. The same trend can be observed in the volatility of these compounds. Whereas the volatility of the trihalides is not very different from that of the corresponding rare-earth halides, the tetrahalides are rather volatile, and the pentahalides, in particular the hexahalides UF_6 and UCl_6, are very volatile.

Interesting is also the trend shown in the stability of the halides, which decreases with increasing atomic number of the halogen

TABLE 12

SOME PROPERTIES OF URANIUM TETRAHALIDES

Compound	M.pt. ($°C$)	B.pt. ($°C$)	Heat of formation $-\Delta H_{298}$ (kcal/mole)
UF_4	1036	1415	450
UCl_4	590	792	251.3
UBr_4	519	761	197.5
UI_4	506	756	126.5

atom. Of the hexahalides, only UF_6 and UCl_6 are known; the existence of UBr_5 being still very problematic. The tetrahalides exhibit a pronounced decrease in stability with increasing atomic number of the halogen atom: for instance UI_4 decomposes into UI_3 and iodine at moderate temperatures. Other properties of interest, such as the melting point, boiling point and stability follow the same trend; this is illustrated in Table 12 for the tetrahalides.

Uranium fluorides

Introduction

In the uranium–fluorine system the compounds UF_3, UF_4, U_4F_{17}, U_2F_9, UF_5 and UF_6 have been found. Their volatility increases from UF_3, which is practically non-volatile, to the highly volatile UF_6. Because the hexafluoride is the only stable, volatile uranium compound, it is of great importance in uranium technology, being used for the separation of the uranium isotopes by gas diffusion. However, for the production of the metal the tetra-fluoride, UF_4, is the important compound.

In the preparation of this group of fluorides, not only fluorine and hydrofluoric acid are employed, but also such compounds as $NH_4F \cdot HF$, CCl_2F_2, CCl_3F and BrF_3 since these may be used at lower temperatures and under milder conditions then those involved in the reaction with fluorine. Thus, parallel with the development of uranium chemistry during the years 1940–1950 there occurred a rapid development in fluorine chemistry generally[1].

Uranium trifluoride

Uranium trifluoride, UF_3, is a dark-violet to black compound with a melting point of about $1425\,°C$[2]. Its crystal structure is hexagonal with lattice parameters $a = 7.179$ Å and $c = 7.345$ Å[3]. The cell contains six molecules and its X-ray density is 8.965.

Preparation. The preparation of UF_3 can best be effected by the reduction of UF_4 with aluminium[4], hydrogen, or uranium metal[5]. In the last instance a uranium–uranium tetrafluoride mixture can be heated in hydrogen at 250 °C to give the finely divided UH_3 powder which permits intimate mixing of hydride with tetrafluoride. The mixture is heated to decompose the hydride and then raised to a temperature of about 1100 °C in argon to complete the reaction.

Chemical properties. Uranium trifluoride is insoluble in water and not very hygroscopic[6]. In water a slow oxidation occurs; this proceeds at 100 °C with a measurable rate. In acids, especially in the cold, the trifluoride resists attack, but it can be converted to UF_4 with HCl probably by the following reaction[5]:

$$4\ UF_3 + 4\ HCl \rightarrow 3\ UF_4 + UCl_4 + 2\ H_2 \uparrow.$$

In oxidizing acids UF_3 is readily converted to a uranyl salt. When heated in air, the trifluoride is oxidized, first to oxyfluoride, and at about 800 °C to U_3O_8. With halogens it reacts at about 250–300 °C to form the corresponding halotrifluoride, for instance with chlorine the compound $UClF_3$[5].

Thermodynamic properties. The heat of formation of solid UF_3 has been reported[7] to be $-\Delta H_{298} = 345$ kcal/mole. This value is consistent with the value estimated by Rand and Kubaschewski[8]. For the free energy of formation it has been found by EMF-measurements[9] $\Delta G_{873}^0 = -310.5$ kcal/mole. Uranium trifluoride disproportionates at 1600 °C into the tetrafluoride and uranium[4]:

$$4\ UF_3 \rightarrow 3\ UF_4 + U$$

The system UF_3–UF_4 has been examined by Long and Thoma[2] who found a eutectic at 892 °C and a solid solubility of UF_4 in UF_3 of 38 mole % at the eutectic.

Uranium tetrafluoride

Uranium tetrafluoride, UF_4, is a green solid ("green salt") which has two polymorphs. Below 833 °C it is monoclinic; the crystal structure has been determined from single-crystal X-ray data[10] to give $a = 12.73$ Å, $b = 10.75$ Å, $c = 8.43$ Å and $\beta = 126° 20'$ (space group C 2/c). The unit cell contains 12 formula units per cell and the X-ray density is 6.70. The uranium atoms are surrounded by eight fluorine atoms arranged in a slightly distorted antiprism configuration. Above 833° the β-modification exists which has a melting point of 1036 °C[11].

Preparation. Uranium tetrafluoride can be prepared either from aqueous solutions or in a dry way. Various modifications of the aqueous preparation have been described. In general a solution of U^{4+} ions, produced by reduction of UO_2^{2+} ions with Sn^{2+} ions or electrochemically, is treated with F^- ions to give a hydrated uranium tetrafluoride. The precipitate is dried and then dehydrated at 400 °C in an atmosphere of dry hydrogen fluoride. It is however essential to prevent hydrolytic reactions during the dehydration, since these lead to a product which is unsuitable either for the production of uranium metal or for conversion to the hexafluoride.

In operations on an industrial scale, the dry procedure is preferred for the tetrafluoride. By this method the oxide UO_2 is converted to UF_4 with hydrogen fluoride at temperatures of 500–700 °C, according to the equilibrium:

$$UO_2 + 4\,HF(g) \rightleftarrows UF_4 + 2\,H_2O(g)$$

Initially, great difficulty was experienced in carrying conversion to completion. However, it was eventually found that UO_2 prepared from ammonium uranate (ADU) was quite satisfactory, whereas that prepared from uranyl nitrate failed to be completely converted. To overcome these difficulties, much work has been done on the kinetics of this reaction. It proved that the rate at which uranium dioxide reacts with HF depends not only on the temperature of hydrofluorination, but markedly on its reactivity (related to specific surface area and particle-size distribution) and on its preparative history (especially the starting material). Thus, samples of UO_2, prepared from ADU and from uranyl nitrate have optimum reaction temperatures which are quite different from each other[12]. This is due to the differences in reactivity between the two UO_2 preparations.

The fluorination is further complicated by loss of surface area, due to crystallite expansion and sintering of the uranium tetrafluoride, resulting in a decrease in reactivity beyond the optimum reaction temperature. The influence of the hydrogen fluoride partial pressure on the reaction has been examined by Tomlinson et al.[13]. In the later stages of the reaction they found the rate to be directly proportional to the partial pressure. The reaction liberates considerable heat (43.2 kcal/mole) so that much care has to be taken to prevent sintering of the particles in the industrial furnaces.

In France, the reaction is carried out in a moving-bed reactor[14], whereas in Great Britain fluid-bed reactors have been developed[15].

It is also possible to employ freons (fluorinated hydrocarbons) in the preparation of UF_4; in these circumstances it is not necessary to start from the tetravalent oxide, since freon itself posesses reducing properties. Attractive is the reaction with CCl_2F_2 which takes place at $400\,°C$[16]:

$$UO_3 + 2\,CCl_2F_2 \rightarrow UF_4 + CO_2 + COCl_2 + Cl_2$$

Provided some oxygen is introduced, phosgene is not formed[17].

Where smaller quantities of UF_4 are required, for instance, for research purposes, alternative reactions may be more attractive. Thus uranium tetrafluoride can be prepared by heating uranium metal with anhydrous, liquid HF in a sealed tube at $\sim 250\,°C$, or by treating UH_3 with gaseous HF at $250–350\,°C$.

Chemical properties. Uranium tetrafluoride is a stable and very unreactive compound. For example, with oxygen it reacts at $800\,°C$, thus[18]:

$$2\,UF_4 + O_2 \rightarrow UF_6 + UO_2F_2$$

With chlorine there is hardly any reaction, but with fluorine it readily forms UF_6 at temperatures above $250\,°C$.

As said before, hydrogen and uranium metal reduce UF_4 to the trifluoride at about $1000\,°C$, whereas the alkaline and earth alkaline metals reduce the tetrafluoride to uranium metal. Industrially, the reaction with magnesium or calcium is used in the preparation of the metal (see Chapter 3).

UF_4 is soluble in ammonium oxalate, but it is almost insoluble in such acids as nitric or hydrochloric acid. Its solubility in water is 0.1 mmole/l[6]. Uranium tetrafluoride is not hygroscopic, but it hydrolizes readily at high temperatures according to the reaction:

$$UF_4 + 2 H_2O \rightleftarrows UO_2 + 4 HF$$

Hydrates. When HF is added to a solution containing U^{4+} ions a hydrated form of UF_4 is precipitated. Below 90 °C it is the gelatinous hydrate $UF_4 \cdot 2\frac{1}{2}H_2O$, at higher temperatures, 95–100 °C, the hydrate $UF_4 \cdot \frac{3}{4}H_2O$. The latter hydrate can be dried in air at 110 °C; it is dark-green and very stable. For the dehydration of the compound a temperature of 400–500 °C is required.

Besides these hydrates two others have been reported: one is a cubic hydrate $UF_4 \cdot 1\frac{1}{2}H_2O$[19] and the other is stated to have the composition $UF_4 \cdot 1 \cdot 4H_2O$[20]. The precise conditions for the formation of these hydrates and their respective stabilities are not known.

Thermodynamic properties of UF_4. The heat of formation of UF_4 is estimated to be $-\Delta H_{298} = 450$ kcal/mole[8]; the entropy as deduced from low-temperature heat-capacity measurements is $S^0_{298} = 36.25$ cal/deg. mole[8]. The melting point of UF_4 is 1036 °C[21]. The vapour pressure of *liquid* UF_4 has been measured by Langer and Blankenship[21], who obtained:

$$\log p(\text{mm}) = \frac{-16,840}{T} - 7.549 \log T + 37.086$$

For *solid* UF_4 the vapour pressure has been evaluated from the above-mentioned vapour pressures for the liquid and the heat of fusion (10.2 kcal/mole) reported by King and Christensen[22] to give[8]:

$$\log p(\text{mm}) = \frac{-16,412}{T} - 3.016 \log T + 22.614$$

The data for solid UF_4 are probably unreliable since it is claimed that they are consistent with a melting point of 960 °C.

Binary systems. The binary systems of UF_4 with U(IV) halides are of interest, especially the UF_4–UCl_4 system which has been examined in detail by Khripin *et al.*[10]. The compounds $UClF_3$, UCl_2F_2 and UCl_3F have been identified and the phase diagram has been determined.

The three chlorofluorides all melt incongruently at 530°, 460° and 444 °C respectively; their thermodynamic properties have been determined by Maslov[23].

The fluorides U_2F_9, U_4F_{17} *and* UF_5

It is convenient to treat these fluorides together since their preparation and properties are closely related. For instance, all can be produced by the reaction of UF_6 with UF_4, or by the reduction of the hexafluoride with hydrogen at 400–600 °C.

The fluoride U_2F_9 is a black solid with a body-centered cubic crystal structure; its lattice parameter is 8.4716 Å and the cell contains four molecules[24]. The U–F distance is 2.30 Å.

The crystal structure of U_4F_{17} is not known.

The compound UF_5 has two modifications. α-UF_5 has a tetragonal cell with $a = 6.512$ Å and $c = 4.463$ Å[24a] and two molecules in the unit cell. The phase β-UF_5 has also a tetragonal cell with $a = 11.45$ Å and $c = 5.198$ Å and 8 molecules in the unit cell.

Thermodynamic properties. Two forms of UF_5 are known of which β-UF_5 exists below about 135 °C and the α-form at higher temperatures (> 150 °C). The transition appears to be rather slow; at 165 °C the conversion requires less than three hours[25].

The vapour pressure of solid and liquid UF_5 have been measured by Wolf *et al.*[26]. They obtained:

for solid UF_5 $\qquad \log p\text{(mm)} = \dfrac{-8001}{T} + 13.994$ $\quad$ (515–619 °K)

for liquid UF_5 $\qquad \log p\text{(mm)} = \dfrac{-5388}{T} + 9.819$ $\quad$ (619–685 °K)

From these measurements it follows that the heat of fusion of UF_5 is 11.9 kcal/mole. The melting point of UF_5 has been given as 346 °C[25] and 348 °C[26]. The melting point determinations have been carried out under a pressure of UF_6 greater than 1.6 atm, since at lower UF_6 pressures U_2F_9 and U_4F_{17} are successively formed. Thus, each of the fluorides UF_5, U_2F_9 and U_4F_{17} disproportionates to the next lowest fluoride and $UF_6(g)$; for instance, UF_5 behaves thus:

$$3 \text{ UF}_5(s) \rightleftarrows U_2F_9(s) + UF_6(g)$$

The disproportionation pressures have been measured by Agron[25], but he does not seem to have corrected his results for the volatility of UF_5.

For the heat of formation of gaseous UF_5 the value $-\Delta H_{298} = 380.8$ kcal/mole has been found[27].

Preparation and chemical properties. The compound UF_5 can be prepared by the reaction of gaseous UF_6 with hydrogen bromide; a reaction which proceeds rapidly at about 65 °C to yield UF_5 of a purity of at least 95 %[28, 29].

$$2 \text{ UF}_6 + 2 \text{ HBr} \rightleftarrows 2 \text{ UF}_5(s) + 2 \text{ HF} + Br_2$$

An excess of UF_6 (molar ratio UF_6/HBr at least 1.7) is necessary to prevent the disproportionation of UF_6 to U_2F_9 or U_4F_{17}. Therefore, in making small amounts of UF_5, a slow stream of HBr is introduced into the apparatus which already contains the hexafluoride. The reaction may be carried out conveniently on a large scale[28].

UF_5 can also be prepared by the reaction of UF_4 with fluorine at 150–250 °C or by the reaction of UF_4 with gaseous UF_6. The latter reaction proceeds very slowly and the type of fluoride formed depends on the temperature, the UF_6 partial pressure and the specific surface area of UF_4. The chemical properties of the intermediate fluorides have not been very well investigated. The compound UF_5 hydrolizes readily in the presence of water; U_2F_9 and U_4F_{17} are reported to be less reactive to water vapour[30].

Uranium hexafluoride

Uranium hexafluoride was discovered in 1909 by Ruff and Heinzelmann[31] who prepared it by the reaction of fluorine with uranium metal or uranium carbide. But the compound received little attention until the second world war. Since that time it has been extensively used for the separation of the uranium isotopes

References p. 162

by gas diffusion, since UF_6 is by far the most volatile uranium compound.

At room temperature, UF_6 is an almost white solid with a remarkably high vapour pressure (112 mm). It has a rhombic crystal structure (space group Pnma) with lattice parameters $a = 9.900$ Å, $b = 8.962$ Å and $c = 5.207$ Å. The unit cell contains four molecules and the X-ray density is 5.09[32].

Preparation. Uranium hexafluoride is prepared on a large scale by the reaction of powdered uranium tetrafluoride with fluorine at about 300 °C:

$$UF_4(s) + F_2 \rightarrow UF_6(g) + 60 \text{ kcal}$$

In the original process by Abelson of the Bureau of Standards (1941), the use of sodium chloride as a catalyst and a temperature of 275 °C was proposed[6]. But shortly thereafter it appeared[33, 34] that a catalyst is not necessary when the temperature is taken somewhat higher. The reaction is more complicated than indicated by the simple equation, since, at temperatures between 250 and 540 °C the intermediate fluorides UF_5, U_2F_9 and U_4F_{17} may also be formed. The yield of UF_6 depends on the temperature, the specific surface area of the UF_4 and the partial pressure of fluorine. Moreover, impurities such as H_2O, $UO_2(U_3O_8)$ reduce the yield[35].

Since 1948 continuous processes for the production of UF_6 have been developed. In general the complete operation involves three steps: (a) the reduction of UO_3 (prepared through either ADU or uranyl nitrate), (b) the hydrofluorination of the resultant UO_2 and (c) the fluorination of UF_4 to produce UF_6. The conversion to UF_6 is essentially complete if a fluorine excess is maintained. The outcoming gas is cooled to about 120 °C by passage through a steam-jacketed pipe and then refrigerated after which UF_6 can be recovered in a solid form.

The cost of producing UF_6 in the USA from crude U_3O_8 concentrates will fall in price from \$1.23/1b U in 1965 to \$1.00 in 1970 and presumably 50¢ in 1975. This is due to process improvements and production on an increasingly large scale. The cheapest process is the fluoride volatility process of the Allied Chemical

Corporation (USA). This process involves the processing of a crude concentrate which has been prepared as a fluidizable powder in fluidized-bed reduction, hydrofluorination and fluorination reactors.

Other processes for the preparation of UF_6—in which the use of elemental fluorine is avoided—have been developed. One of them, the *fluorox* process, is of interest since it works with the relatively cheap HF. In this process, uranium tetrafluoride is oxidized with oxygen and the non-volatile UO_2F_2 which is a byproduct, is recycled through UO_2 to give UF_4:

$$UF_4 + O_2 \xrightarrow{\ 800°C\ } UF_6(g) + UO_2F_2$$

The kinetics of the reaction between oxygen and UF_4 have been investigated by Ferris[36] who found the rate to be first order with respect to the UF_4 surface area.

The fluorox process has been investigated in a fluidized bed plant at the Oak Ridge National Laboratory, USA[37]. The formation of UO_2F_2 is a disadvantage since it gives a lower yield of UF_6 than does the direct fluorination process, and it necessitates a reprocessing stage to convert UO_2F_2 to UF_4. This is generally done by reducing UO_2F_2 to UO_2 and hydrofluorinating the oxide to UF_4.

A further disadvantage of this process is that corrosion in the fluorox reactor is higher than in the fluorination reactor, mainly since higher reaction temperatures are used. Moreover, UF_4 sinters appreciably at 800 °C which causes inadequate fluidization and mechanical blockages. For this reasons different types of reactors has been studied, for instance, the type which employs a moving bed.

Chemical properties. Uranium hexafluoride is not a very reactive substance. Under ordinary conditions it does not react with oxygen, chlorine, or nitrogen. The reaction with hydrogen proceeds only above 300 °C and is very slow, even at 600 °C.

However, UF_6 reacts violently with water to form uranyl fluoride, UO_2F_2, and HF, liberating considerable heat. Uranium hexafluoride is reduced to the tetrafluoride when heated in hydrogen, but the reaction has a high energy of activation, and even at 600 °C proceeds only slowly. It is more readily reduced to UF_4 by hydrogen chloride (at 250 °C) and by hydrogen bromide (at 80 °C). Ammonia reacts rapidly even at -78 °C to give NH_4UF_5.

Towards organic substances, such as alcohol, ether, benzene or hydrocarbons, UF_6 behaves as a powerful fluorinating agent, yielding in the process HF and either UO_2F_2 or UF_4.

The phase diagram and thermodynamic properties. The vapour pressure of solid and liquid UF_6 has been measured many times with results in excellent agreement. A review of these has been given by DeWitt[38] and smoothed values are given in Table 13. It follows from the measurements that the sublimation temperature of UF_6 is 56.4 °C and the triple point 64.02 °C (at 1137.5 mm). For the critical point, it has been found $t_c = 230.2$ °C and $p_c = 45.5$ atm[39]. The phase diagram is shown in Fig. 29.

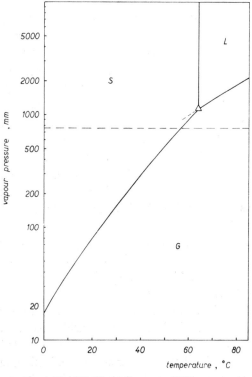

Fig. 29. Phase diagram of UF_6.

TABLE 13

VAPOUR PRESSURE OF URANIUM HEXAFLUORIDE

Temperature, °C	Pressure, mm	
−10	7.7	
0	17.6	
10	40	
20	79	
25	112	
30	154	solid
40	295.5	
50	523	
56.4 (subl. pt.)	760	
60	920	
64.02 (triple pt.)	1137.5	
70	1350	
80	1840	liquid
90	2400	
100	3000	

The heat of formation of solid UF_6 from the elements has been measured directly in a bomb calorimeter[27]. It has been found $-\Delta H_{298} = 522.6 \pm 0.4$ kcal/mole. For gaseous UF_6, $-\Delta H_{298} = 510.8$ kcal/mole and from which the heat of sublimation is 11.8 kcal/mole. This value is in excellent agreement with the values found by Masi[40] calorimetrically.

The standard entropy of solid UF_6, from the low-temperature heat capacity measurements[6], is $S_{298}^0 = 54.45$ cal/deg. mole. The heat capacity of solid and liquid UF_6 has been measured from 14–370 °K[41] and from 210–273 °K[42], whereas the heat capacity of the vapour has been calculated from spectroscopic data[43,44]. For the entropy of gaseous UF_6 it has been found $S_{298}^0 = 90.34$ cal/deg. mole, which is in nice agreement with the value (90.4 kcal) found from the entropy of the solid and the entropy of sublimation.

The melting-point diagram for the UF_6–PuF_6 system has been examined recently[44a] and a continuous series of solid solutions has been found.

Physical properties. Uranium hexafluoride is paramagnetic; its molecular susceptibility is reported to be $+106 \cdot 10^{-6}$ cgs units at

300 °K and to be independent of temperature[45]. The dipole moment of UF_6(g) is essentially zero, and thus the UF_6 molecule may be considered as a regular octahedron. This is in agreement with deductions from infrared and Raman spectrum studies[43,44,46].

The thermal conductivity of the vapour between 0 and 100 °C can be represented by:

$$K = 1.46(1+0.0042 \cdot t) \cdot 10^{-5}$$

in which K is in cal/cm.sec.deg and t in °C.

The density of solid UF_6 is 5.06 at 25 °C, calculated from the X-ray lattice constants. The liquid density has been measured as a function of temperature; at 65.1 °C it was 3.667^{47}, whereas at the triple point (64.02 °C) the value was 3.674^{41}.

In disagreement with earlier findings, recent measurements have shown that the vapour of UF_6 does not behave as an ideal gas. Between 50–140 °C the following equation of state has been reported[48]:

$$\varrho = 4.291 (P/T) \cdot (1 + 1.2328 \cdot 10^6 \; P/T^3)$$

Uranium nitrogen fluoride, UNF, see p. 198.

Uranium chlorides

The uranium chlorides UCl_3, UCl_4, UCl_5 and UCl_6 have been described. Because of the larger radius of the Cl^- ion (1.81 Å, compared with 1.36 Å for the F^- ion) the chlorides have lower melting and boiling points than the corresponding fluorides.

The trend in physical properties, such as melting point and boiling point, density and heat of formation, however, is less pronounced than it is for the fluorides. The same holds for the chemical properties. For instance, reactivity to water vapour is high for all chlorides, in distinction from the fluorides of which only UF_6 is very reactive.

Uranium trichloride

At room temperature UCl_3 is an olive-green substance; at

higher temperatures it reversibly darkens in colour going through red to dark-purple at 450 °C. The melting point is 842 °C, and the crystal structure is hexagonal (space group $P6_3/m$) with $a = 7.441$ Å and $c = 4.322$ Å. It has two molecules in the unit cell[49] and has a density of 5.51.

Preparation. Uranium trichloride can be prepared by the reaction of hydrogen chloride with uranium hydride at 250–300 °C:

$$UH_3 + 3\ HCl \rightarrow UCl_3 + 1\tfrac{1}{2}\ H_2$$

However, the hydrogen chloride is strongly absorbed by UCl_3 and can only be removed in a vacuum at 150 °C. Moreover, the product is mostly contaminated with UCl_4.

Other methods for the preparation of UCl_3 consist of reducing UCl_4 with hydrogen or zinc metal[6]:

$$2\ UCl_4 + Zn \xrightarrow{450\,°C} 2\ UCl_3 + ZnCl_2$$

UCl_3 can be purified by distillation with iodine at 500 °C. The reduction of UCl_4 with hydrogen is very slow below its melting point (590 °C). At higher temperatures the surface of the UCl_4 becomes covered with a layer of UCl_3 which prevents further reduction.

Probably the best procedure for the preparation of UCl_3 is the reduction of UCl_4 with hydrogen in the apparatus, shown in Fig. 30[50]. Hydrogen is passed into the apparatus through tube a. The temperature of the UCl_4 on the coarse-porosity fritted-glass disk b, which is measured with the thermocouple inside the thermocouple well c, is raised to 500 °C and maintained at this value for six hours. It is then raised to, and held at, 540 °C for 16 hours. Finally, the temperature is maintained at 580 °C for one hour, after which the furnace is allowed to cool. Initial reduction at lower temperatures is necessary to prevent formation of a sintered product that may contain unreacted and occluded UCl_4.

Thermodynamic properties. The heat of formation of UCl_3 has been determined from heat of solution measurements in $HCl/FeCl_3$ mixtures[8]. It has been found that $-\Delta H_{298} = 213.5$ kcal/mole. The entropy of UCl_3 as deduced from low-temperature heat-capac-

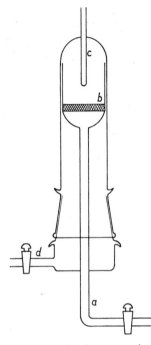

Fig. 30. Apparatus for the preparation of UCl_3.

ity measurements[6] is $S^0_{298} = 38.0$ cal/deg. mole. High-temperature heat-capacity measurements have also been made [50a]; the results can be represented by:

$$c_p = 20.8 + 7.75 \cdot 10^{-3} \, T + 1.05 \cdot 10^5 \, T^{-2}$$

The thermal decomposition of UCl_3, according to:

$$4 \, UCl_3(s) \rightarrow 3 \, UCl_4(l) + U(s)$$

has been examined by Hardy-Grena[51], who measured vapour pressures in the temperature range 750–900 °C.

Chemical properties. UCl_3 is hygroscopic but to a much smaller degree than the other uranium chlorides. It is a strong reducing agent, both as a solid and in solution. In water, UCl_3 dissolves readily to give a purple solution which, with the evolution of hydrogen, soon turns dark green owing to the formation of

U^{4+} ions. The oxidation has been followed by observing changes in the absorption spectra[52]. With ammonia UCl_3 undergoes ammonolysis above 400°C[53]; NH_4Cl and amorphous amido-chlorides are formed and these are transformed at about 800°C, into a uranium nitride with the approximate composition $UN_{1.7}$.

Uranium trichloride is insoluble in non-polar solvents, such as CCl_4, acetone and chloroform. With chlorine, it reacts at 250°C to form UCl_4; with oxygen reaction begins at 150°C, but in air at a much higher temperature. The product is uranyl chloride, UO_2Cl_2.

Uranium tetrachloride

Uranium tetrachloride, UCl_4, is a dark-green solid melting at 590°C. It has a tetragonal crystal structure (space group 14/amd) with $a = 8.298$ Å and $c = 7.486$ Å and four molecules in the unit cell[54]. A phase transition in UCl_4 at 547°C has been reported[20].

Preparation. There are a number of reactions by which UCl_4 can be prepared[6]. In general, it can be made by passing a chlorine-containing gas (Cl_2, CCl_4 or $SOCl_2$) over uranium, uranium hydride or a mixture of uranium oxide and carbon at temperatures of 500–700°C.

The direct combination of uranium with chlorine gives a complex mixture of uranium chlorides and, accordingly, does not provide a useful preparative method. To convert UO_2 to UCl_4 a powerful halogenation reagent is required. Carbon tetrachloride CCl_4, and especially thionyl chloride, $SOCl_2$, are very effective; they convert UO_3 (or U_3O_8) rapidly to UCl_4 at 350–400°C. However, owing to the higher valency of uranium in UO_3, chlorine is formed and, as a result, comparatively large amounts of uranium pentachloride are produced which are swept out as dust in the gas stream:

$$UO_3 + SOCl_2 \rightarrow UCl_4 + SO_2 + Cl_2$$
$$2\,UCl_4 + Cl_2 \rightarrow 2\,UCl_5$$

When UO_2 is used, the principal product with thionyl chloride or CCl_4 at 450°C is the tetrachloride:

$$UO_2 + CCl_4 \rightarrow UCl_4 + CO_2$$

Small amounts of CO, phosgene and chlorine are always found in the exit gases. Uranium tetrachloride can be purified by sublimation in argon at a temperature of 700–900 °C[55].

UCl$_4$ can also be prepared from uranium pentachloride by heating it in vacuum or in an inert gas at 150 °C, or by reducing it in hydrogen at about 350 °C. For the preparation of UCl$_4$ in the liquid phase, the reader is referred to the book of Katz and Rabinowitch[6].

Thermodynamic properties. For the heat of formation of UCl$_4$ the value $-\Delta H_{298} = 251.3$ kcal/mole has been found[8]. The entropy of solid UCl$_4$ at 298 °K is 47.14 cal/deg.mole[6]. The high-temperature heat capacities of UCl$_4$ have also been measured[6] and the results can be summarized as follows:

$$C_p = 27.2 + 8.57 \cdot 10^{-3} T - 0.79 \cdot 10^5 T^{-2} \ (298\text{--}800\,°K)$$

The vapour pressure of both solid and liquid UCl$_4$ have been measured independently by a number of authors but the various results do not agree very well. The most reliable values are probably those by Mueller[56], who found for solid UCl$_4$:

$$\log p(\text{mm}) = \frac{-10{,}427}{T} + 13.2995 \quad (623\text{--}778\,°K)$$

For the vapour pressure of liquid UCl$_4$ it has been found[53, 54]:

$$\log p(\text{mm}) = \frac{-7205}{T} + 9.65 \quad (863\text{--}1063\,°K)$$

According to the measurements of Mueller[56], the triple point of UCl$_4$ is at 590 °C and 19.5 mm.

Chemical properties. Uranium tetrachloride is very reactive to water vapour; it readily hydrolizes to uranium(IV) oxychloride, UOCl$_2$. In water it dissolves with the evolution of heat; the solution is green due to the presence of U^{4+} ions which are strongly hydrolized. The hydrate UCl$_4 \cdot 9H_2O$ is reported to crystallize from saturated solutions of uranium tetrachloride[58].

Uranium tetrachloride is readily soluble in polar solvents, but in non-polar solvents, such as benzene, ether, or hydrocarbons, it is insoluble. UCl_4 is soluble in acetone, probably with chemical reaction. UCl_4 oxidizes in air at about 300 °C to form uranyl chloride, UO_2Cl_2 (see p. 131); at higher temperatures U_3O_8 is formed. Hydrogen reduces UCl_4 at 500–550 °C to give UCl_3.

With ammonia, UCl_4 undergoes ammonolysis at 450 °C to form uranium(IV) amidotrichloride, UNH_2Cl_3[53]. At about 475 °C this compound is reduced to UCl_3; the ammonolysis then continues as for UCl_3.

Uranium pentachloride

The pentachloride is a red-brown substance. It is often obtained as a by-product in the preparation of UCl_4, as a volatile dust. UCl_5 has a monoclinic cell (space group $P2_1/n$) with lattice parameters $a = 7.99$ Å, $b = 10.69$ Å, $c = 8.48$ Å and $\beta = 91.5°$; the X-ray density is 3.81[59]. In this structure, each uranium atom is coordinated octahedrally to six chlorine atoms; two octahedra share an edge to form a U_2Cl_{10} unit of which there are two dimers per unit cell. The presence of U_2Cl_{10} dimers has also been observed in solutions prepared from CCl_4[60].

Preparation. Uranium pentachloride can be prepared (a) by chlorination of UCl_3 or UCl_4 at 500–550 °C, for instance with chlorine or CCl_4, (b) by liquid phase reaction of uranium oxides with UCl_4. For instance, UO_3 can be converted to UCl_5 at the boiling point of CCl_4; a large excess of CCl_4 has to be refluxed for several hours. The reaction is better be carried out under pressure; for instance in a sealed tube filled with CCl_4 and heated overnight at about 225 °C. Afterwards the impurities of UCl_3 or UCl_4, which are insoluble in CCl_4, are filtered off at 100 °C, and the solution is slowly cooled to room temperature. It yields large crystals of UCl_5[60].

Thermodynamic properties. The heat of formation of UCl_5 is reported to be $-\Delta H_{298} = 261.5$ kcal/mole; the entropy has been estimated to be 58 cal/deg.mole[8].

UCl_5 is not very stable; when heated at about 150 °C it both decomposes and disproportionates:

$$2\ UCl_5 \nearrow^{2\ UCl_4 + Cl_2}_{\searrow UCl_4 + UCl_6(g)}$$

For this reason reliable data on the high-temperature properties of the compound, such as the volatility, are not available.

Chemical properties. Uranium pentachloride is very hygroscopic; in water it decomposes immediately according to:

$$2\ UCl_5 + 2\ H_2O \rightarrow UCl_4 + UO_2Cl_2 + 4\ HCl$$

UCl_5 is soluble in CCl_4, CS_2 and $SOCl_2$, which is in contrast to UCl_3 and UCl_4. With oxygen-containing organic solvents, such as acetone or alcohols, it reacts immediately.

Uranium hexachloride

Uranium hexachloride is an interesting compound since it is, besides UF_6, the only known hexavalent uranium compound which does not contain oxygen. It has an unexpected stability and a marked volatility. The black to dark-green substance has a hexagonal crystal symmetry; the lattice parameters of the unit cell are $a = 10.90$ Å and $c = 6.03$ Å. The unit cell contains three molecules and the X-ray density is 3.59[61].

Preparation. Uranium hexachloride may be prepared by the disproportionation of UCl_5 in a high vacuum at 125–150 °C:

$$2\ UCl_5 \rightarrow UCl_4 + UCl_6$$

The volatile hexachloride is collected on a cold finger; a special adaption of the apparatus has been described by Johnson *et al.*[62].

Another method consists in the chlorination of lower chlorides to UCl_6 by means of chlorine above 350 °C. The impure product contains about 30 % UCl_6 and can be purified by sublimation in a high vacuum at about 100 °C.

Thermodynamic properties. The heat of formation of UCl_6 has been found to be $-\Delta H_{298} = 270.7$ kcal/mole; the entropy is reported to be 68.3 cal/deg.mole[6].

The melting point of UCl_6 is $177.5 \pm 2.5\,°C$. For the vapour pressure it has been found[62]:

$$\log p(\text{mm}) = \frac{-3788}{T} + 9.52$$

These pressures are, however, of rather uncertain validity, since some decomposition of UCl_6 has been observed at the measuring temperatures, probably according to:

$$2\,UCl_6 \rightarrow 2\,UCl_5 + Cl_2$$

Chemical properties. Uranium hexachloride reacts violently with water to form uranyl chloride, UO_2Cl_2. With HF it gives UF_5, hydrogen and chlorine. In CCl_4 the UCl_6 dissolves to give a stable brown-coloured solution.

Uranium bromides and iodides

The uranium bromides and iodides which are definitely known are limited to those with the compositions UX_3 and UX_4. They resemble the corresponding fluorides and chlorides in many respects.

Uranium tribromide

Uranium tribromide is a dark-brown substance with a hexagonal crystal structure. The unit cell contains two molecules and has the dimensions $a = 7.926$ Å and $c = 4.432$ Å. For the melting point values of $730-755\,°C$ have been reported.

Preparation. Probably the best method of making UBr_3 employs the reaction of uranium hydride with HBr at $300\,°C$. The tribromide can also be made from uranium metal and bromine vapour at $300-500\,°C$. The reduction of UBr_4 with hydrogen is reported to be less satisfactory.

Properties. Uranium tribromide is very hygroscopic, much more so than the corresponding chloride. It dissolves in water, evolving hydrogen, and it reacts with chlorine to give uranium tetrachloride and with bromine to give the tetrabromide. Oxygen reacts with UBr_3 even at room temperature. UBr_3 is almost insoluble in non-polar solvents.

Thermodynamic properties. For the heat of formation of UBr_3 the value $-\Delta H_{298} = 172.3$ kcal/mole has been found[8]. Vapour pressures have been measured and the results can be given as:

$$\log p(\text{mm}) = \frac{-15,000}{T} + 12.50 \quad (900\text{–}1250\,^\circ\text{K})$$

The phase diagram of the UBr_3–UBr_4 system has been examined[63]. It proves to be a simple eutectic one with negligible solubility in the liquid phase, but with an appreciable solubility of UBr_3 in UBr_4. The eutectic is at $490\,^\circ$C and 76 mole per cent UBr_3 (Fig. 31)

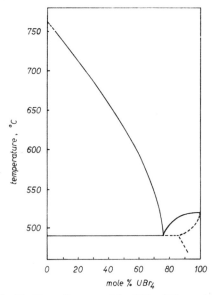

Fig. 31. Phase diagram of the UBr_3–UBr_4 system.

Uranium tetrabromide

Uranium tetrabromide resembles uranium tetrachloride quite closely in its properties, which, however, have been much less extensively investigated. It is a brown substance with a melting point of 519 °C. The crystals have a monoclinic structure with lattice parameters $a = 10.92$ Å, $b = 8.69$ Å, $c = 7.05$ Å and $\beta = 93.9°$. The X-ray density is 5.53[64].

Preparation. Uranium tetrabromide can be prepared by heating bromine with uranium metal or with other uranium compounds, such as UH_3 or $UO_2(+C)$. It can also be made by heating UO_3 at 165 °C with carbon tetrabromide, CBr_4. A convenient method is the reaction of bromine with uranium tribromide at 300 °C. Uranium tetrabromide can be purified by distillation in nitrogen at 600 °C.

Thermodynamic properties. The heat of formation of UBr_4 has been found to be $-\Delta H_{298} = 197.5$ kcal/mole. Vapour pressure measurements have been carried out[63] and the results are:

for solid UBr_4 $\qquad \log p(\text{mm}) = \dfrac{-10,900}{T} + 14.56$

and for liquid UBr_4 $\qquad \log p(\text{mm}) = \dfrac{-7060}{T} + 9.71$

The boiling point of uranium tetrabromide is 761 °C, but at this temperature a slight decomposition to UBr_3 has been observed[6].

Chemical properties. Uranium tetrabromide is soluble in organic liquids. It is also highly soluble in water, giving a green solution in which it is strongly hydrolyzed.

UBr_4 reacts with chlorine to give uranium tetrachloride and with oxygen to give uranyl bromide. It can be reduced with hydrogen at 500–700 °C to uranium tribromide. With calcium or magnesium reduction to uranium metal occurs.

Uranium pentabromide

The preparation of uranium pentabromide, UBr_5, has been reported by Prigent[65,66]. Recently, a simple method for its preparation has been reported by Lux[66a]. The compound can be obtained by treating UBr_4 or uranium metal with boiling bromine; it is a black powder which is stable to dry oxygen, but very sensitive to moisture.

Uranium iodides

Uranium triiodide and tetraiodide are black, needle-like, crystalline substances, both of which can be prepared by direct combination of the elements. For the preparation of UI_3, the stoichiometric quantity of iodine is slowly distilled into an evacuated tube containing finely divided uranium (obtained from UH_3) at 300 °C. After being sealed off, the mixture is heated for about 20 hours at 550 °C to yield practically pure UI_3. Uranium triiodide can also be prepared by action of hydrogen iodide on uranium hydride.

For the preparation of UI_4 the uranium may be heated with iodine as before, but it is now necessary that the partial pressure of iodine is high enough to prevent the equilibrium:

$$UI_4 \rightleftarrows UI_3 + \tfrac{1}{2} I_2$$

from allowing appreciable amounts of UI_3 to remain. An iodine pressure of about 200 mm is sufficient for this purpose. An adequate procedure for the preparation of UI_4 consists of the iodination of uranium triiodide.

Crystallographic properties. Uranium triiodide has an orthorhombic crystal structure (space group Ccmm) with lattice parameters $a = 13.98$ Å, $b = 4.33$ Å and $c = 9.99$ Å. The unit cell contains four molecules and the X-ray density is 6.76.

Crystallographic data for UI_4 have not been reported.

Thermodynamic properties. The heats of formation $(-\Delta H_{298})$ of UI_3 and UI_4 are 114.2 and 126.5 kcal/mole respectively. For the standard entropy (S^0_{298}) the values 57 and 67 cal/deg. mole have been estimated[8]. For the free energy of formation of UI_3 the value $\Delta G^0_{668} = -98.8$ kcal/mole has been found[67]. The vapour pressure of UI_4 has been measured and yields[6]:

$$\log p(\text{mm}) = \frac{-11,520}{T} + 15.53.$$

UI_4 melts at 506 °C and boils at 756 °C.

UI_3 melts at 766 °C, and it is less volatile than UI_4 but quantitative data are not available.

Chemical properties. The uranium iodides are very hygroscopic;

UI_3 dissolves violently in water to give a dark-red solution. Solutions of UI_4 in water are strongly hydrolyzed.

The uranium iodides are easily oxidized in air. It has been suggested[6] that uranyl iodide, UO_2I_2, is formed at room temperature, but this compound is in itself unstable with respect to evolution of iodine. At slightly higher temperatures the oxidation product is U_3O_8[68]. UI_4 is easily reduced to UI_3 by hydrogen at moderate temperatures.

Uranium oxyhalides

Oxyhalides of both uranium(IV) and uranium(VI) have been well established. In addition, oxyhalides of pentavalent uranium, such as UO_2Cl have been described[69].

The uranyl halides are the oxyhalides of hexavalent uranium; they can be considered as commonly occurring uranium salts and, as such, they have been discussed in Chapter 8. For this reason attention will be limited here only to the uranium(IV) oxyhalides of which the chloride and bromide alone are reported to exist.

Uranium(IV) oxydichloride

Uranium oxydichloride, $UOCl_2$, is a yellow-green substance, that, according to early observations[6], can be obtained by the evaporation of an aqueous solution of uranium tetrachloride. After drying at 100 °C it has the approximate composition $UOCl_2 \cdot H_2O$.

Anhydrous $UOCl_2$ can be prepared conveniently by dissolving UO_2 in an excess of molten uranium tetrachloride at 600 °C. It has an orthorhombic crystal structure (space-group Pbam) with lattice parameters $a = 15.255$ Å, $b = 17.828$ Å and $c = 3.992$ Å[70]. The infrared spectrum of $UOCl_2$ has been examined by Bagnall et al.[70] who showed the absence of discrete UO^+ groups in the oxychloride.

Thermodynamic and chemical properties. The heat of formation of $UOCl_2$ has been estimated to be $-\Delta H_{298} = 260 \pm 2$ kcal/mole[8]. The standard entropy has been measured by Westrum et al.[71], who obtained $S^0_{298} = 33.1$ cal/deg. mole.

The oxydichloride readily dissolves in water to give a green solution. With CCl_4 it reacts at 170 °C to produce UCl_4 and by hydrogen it is reduced, presumably to $UOCl$.

Uranium(V) oxytrichloride

$UOCl_3$ is a reddish-brown substance. It is obtained when stoichiometric amounts of uranyl chloride and uranium tetrachloride are heated together at 370 °C[72, 73].

There is some spectral evidence for the existence of the dioxymonochloride, UO_2Cl, in fused lithium chloride, potassium chloride solutions of uranyl chloride[73].

Uranium(IV) oxybromide

Uranium(IV) oxybromide, $UOBr_2$, is a greenish-yellow substance which is stable and non-volatile at 600 °C in an inert atmosphere. At 800 °C, however, it disproportionates to UO_2 and UBr_4:

$$2 UOBr_2 \rightarrow UBr_4 + UO_2$$

When heated in air it forms U_3O_8. Its heat of formation is estimated to be $-\Delta H_{298} = 240$ kcal/mole[8].

REFERENCES

1 CH. SLESSER AND S. R. SCHRAM, *Preparation, Properties and Technology of Fluorine and Organic Fluoro Compounds*, McGraw-Hill Book Company, Inc., New York (1951).
2 G. LONG AND R. E. THOMA, *ORNL-Report* 3789 (1965).
3 E. STARITZKY AND R. M. DOUGLASS, *Anal. Chem.*, 28 (1956) 1056.
4 R. A. SATTEN, C. L. SCHREIBER AND E. Y. WONG, *J. Chem. Phys.*, 42 (1965) 162.
5 J. C. WARF, in: *Chemistry of Uranium*, Collected papers edited by J. J. KATZ AND E. RABINOWITCH, *TID-Report* 5290 (1958), p. 81.
6 J. J. KATZ AND E. RABINOWITCH, *The Chemistry of Uranium*, McGraw-Hill Book Company, Inc., New-York (1951).
7 J. L. SETTLE AND H. M. FEDER, *J. Phys. Chem.*, 67 (1963) 1892.
8 M. H. RAND AND O. KUBASCHEWSKI, *Thermochemical Properties of Uranium Compounds*, Oliver & Boyd, Edinburgh (1963).
9 R. J. HEUS AND J. J. EGAN, *Z. Phys. Chem.*, 49 (1966)38.
10 A. C. LARSON, R. B. ROOF, JR. AND D. T. CROMER, *Acta Cryst.*, 17 (1964) 555.

11 L. A. KHRIPIN, YU. V. GAGARINSKII, G. M. ZADNEPROVSKII AND L. A. LUK'YANOVA, *At. Energ.* (*USSR*), 19 (1965) 437.

12 B. A. LISTER AND G. M. GILLIES, *Process Chemistry, Progress in Nuclear Energy, Series II*, Vol. I, McGraw-Hill Book Company Inc., New York (1956), p. 19.

13 L. TOMLINSON, S. A. MORROW AND S. GRAVES, *Trans. Faraday Soc.*, 57 (1964) 1008.

14 J. H. GITTUS, *Uranium*, Butterworth, London (1963), p. 123.

15 E. HAWTHORN, *Trans. Inst. Chem. Eng.*, London, 38 (1960) 197.

16 H. S. BOOTH, W. KRASNY-ERGEN AND R. F. HEATH, *J. Am. Chem. Soc.*, 68 (1946) 1969.

17 A. CACCIARI, C. FIZZOTTI, M. GABAGLIO AND R. DELEONE, *Energie Nucléaire*, 1 (1957) 11.

18 S. FRIED AND N. R. DAVIDSON, ref. 5, p. 688.

19 J. DAWSON, R. D'EYE AND A. TRUSWELL, *J. Chem. Soc.*, (1954) 3922.

20 YU. V. GAGARINSKII, E. J. KHANAEV, N. P. GALKIN, L. A. ANAN'EVA AND S. P. GABUDA, *At. Energ.* (*USSR*), 18 (1965) 40.

21 S. LANGER AND F. F. BLANKENSHIP, *J. Inorg. Nucl. Chem.*, 14 (1960) 26.

22 E. G. KING AND A. U. CHRISTENSEN, *BM-RI-Report* 5709 (1961).

23 P. G. MASLOV, *Russ. J. Inorg. Chem.*, 9 (1964) 1122.

24 J. LAVEISSIÈRE, *Bull. Soc. Fr. Miner. Crist.*, 90 (1967) 308.

24a W. H. ZACHARIASEN, *Acta Cryst.*, 1 (1948) 277.

25 P. A. AGRON, ref. 5, p. 610 (Vol. II).

26 A. S. WOLF, J. C. PROSEY AND K. E. RAPP, *Inorg. Chem.*, 4 (1965) 751.

27 P. A. G. O'HARE AND W. N. HUBBARD, *Trans. Faraday Soc.*, 62 (1966) 2709.

28 U. S. Patent 3,035,894 (May 22, 1962).

29 A. S. WOLF, W. E. HOBBS AND K. E. RAPP, *Inorg. Chem.*, 4 (1965) 755.

30 Ref. 6, p. 392.

31 O. RUFF AND A. HEINZELMANN, *Ber.*, 42 (1909) 495.

32 J. L. HOARD AND J. D. STROUPE, ref. 5, p. 325.

33 J. E. VANCE, *National Nuclear Energy Series*, VII, McGraw-Hill Book Company Inc., New York (1951).

34 S. H. SMILEY AND D. C. BRATER, *Proceedings of the Second International Conference on the Peaceful Uses of Atomic Energy, Geneva (1965)*. United Nations, Vol. 4 (1958), p. 153.

35 A. LEVEL, *Energ. Nucléaire*, 4 (1962) 278.

36 C. D. SCOTT, *ORNL-Report* 2797 (1960).

37 L. M. FERRIS, *J. Am. Chem. Soc.*, 79 (1957) 5419.

38 R. DEWITT, *GAT-Report* 280 (1960), p. 81.

39 G. OLIVER, H. MILTON AND I. GRISARD, *J. Am. Chem. Soc.*, 75 (1953) 2827.

40 J. F. MASI, *J. Chem. Phys.*, 17 (1949) 755.

41 F. G. BRICKWEDDE, H. J. HOGE AND R. B. SCOTT, *J. Chem. Phys.*, 16 (1948) 429.

42 D. R. LLEWELLYN, *J. Chem. Soc.*, (1953) 28.

43 J. BIGELEISEN, M. G. MAYER, P. C. STEVENSON AND J. TURKEVICH, *J. Chem. Phys.*, 16 (1948) 442.

44 J. GAUNT, *Trans. Faraday Soc.*, 49 (1953) 1122.

44a L. E. TREVORROW, M. J. STEINDLER, D. V. STEIDL AND J. T. SAVAGE, *Inorg. Chem.*, 6 (1967) 1060.

45 W. TILK AND W. KLEMM, *Z. Anorg. Allgem. Chem.*, 240 (1939) 355.

46 H. H. CLAASSEN, B. WEINSTOCK AND J. G. MALM, *J. Chem. Phys.*, 25 (1956) 426.

47 H. F. PRIEST AND G. L. PRIEST, ref. 5, p. 734.

48 D. W. MAGNUSON, *J. Chem. Phys.*, 24 (1955) 344.

49 E. STARITZKY, *Anal. Chem.*, 28 (1956) 1055.

50 *Inorganic Syntheses*, TH. MOELLER, editor, McGraw-Hill Book Company, Inc. New York. Vol. V (1957), p. 145.

50a D. C. GINNINGS AND R. J. CORRUCCINI, *J. Res. Nat. Bur. Standards*, 39 (1947) 309.

51 C. HARDY-GRENA, *Eur.* 2189 f (1964).

52 J. J. HOWLAND, JR. in: *Chemistry of Uranium*, collected papers edited by J. J. KATZ AND E. RABINOWITCH, T.I.D.-Report 5290, Book 2 (1958), p. 680.

53 H. J. BERTHOLD AND H. KNECHT, *Angew. Chem. (Internat. Ed.)*, 4 (1965) 433.

54 E. STARITZKY, Anal. Chem., 28 (1956) 1056; cf. R. L. L. MOONEY, *Acta Cryst.*, 2 (1949) 189.

55 J. D. HEFLEY, D. M. MATHEWS AND E. S. AMIS, *J Inorg. Nucl. Chem.*, 12 (1959) 84.

56 M. E. MUELLER, *AECD-Report* 2029 (1948).

57 H. S. YOUNG AND H. F. GRADY, *TID-Report* 5290 (1949).

58 A. CHRÉTIEN AND C. POMMIER, *Compt. Rend. [C]*, 262 (1966) 644.

59 G. S. SMITH, Q. JOHNSON AND R. E. ELSON, *Acta Cryst.*, 22 (1967) 300.

60 H. L. GOREN, R. S. LOWRIE AND J. V. HUBBARD, quoted in ref. 6, p. 493.

61 W. H. ZACHARIASEN, *Acta Cryst.*, 1 (1948) 285.

62 O. JOHNSON, T. BUTLER AND A. S. NEWTON, in: *Chemistry of Uranium*, collected papers edited by J. J. KATZ AND E. RABINOWITCH, *TID-Report* 5290, Book 1 (1958), p. 18.

63 F. H. SPEDDING, A. S. NEWTON, R. NOTTORF, J. POWELL AND V. CALKINS, in: *ibid.*, p. 91.

64 R. M. DOUGLASS AND E. STARITZKY, *Anal. Chem.*, 29 (1957) 459.

65 J. PRIGENT, *Compt. Rend.*, 239 (1954) 424.

66 J. PRIGENT, *Ann. Chim.*, 5 (1960) 65.

66a F. LUX (1969); see *Angew. Chem., Nachr. Chem. Techn.*, 17 (1969) 129.

67 J. O. TVEEKREM AND M. S. CHANDRASEKHARAIAH, *J. Electrochem. Soc.*, 115 (1968) 1021.

68 M. HANDA, *Bull. Chem. Soc. Japan*, 39 (1966) 2315.

69 J. C. LEVET, *Compt. Rend.*, 260 (1965) 4775.

70 K. W. BAGNALL, D. BROWN AND J. F. EASEY, *J. Chem. Soc. (A)* (1968) 288.

71 E. GREENBERG AND E. F. WESTRUM, JR., *J. Am. Chem. Soc.*, 78 (1956) 5144.

72 S. A. SHCHUKAREV, I. V. VASIL'KOVA, N. S. MARTYNOVA AND YU. G. MAL'TSEV, *Zhur. Neorg. Khim.*, 3 (1958) 2647.

73 M. D. ADAMS, D. A. WENZ AND R. K. STEUNENBERG, *J. Phys. Chem.*, 67 (1963) 1939.

CHAPTER 10

Compounds of Uranium with Elements of Group IV

Introduction

The compounds of uranium with the Group IVA elements, in particular the carbides, have a unique combination of metallic transport properties with very high melting points and hardness. These properties have made their application as nuclear fuels attractive. The monocarbide UC and a dispersion of the carbide UC_2 in graphite are potentially useful fuels for high-temperature reactors. Incidentally the silicide U_3Si is now being developed as a nuclear fuel. These applications have stimulated an enormous amount of research and development work of which the results have become available during the past five years.

Uranium carbides

A carbide of uranium was first prepared in 1896 by Moissan[1] by the reaction of U_3O_8 with carbon in an electric furnace. He obtained an impure product to which he ascribed the formula U_2C_3. Litz et al.[2] reported a new carbide of uranium, UC, in a lecture before the American Chemical Society in 1946. Shortly before the lecture appeared in print, Rundle et al.[3] published their work on the uranium carbides done in 1942 as part of the Manhatten Project. They had found the carbides UC and UC_2 and secured X-ray evidence of a third carbide U_2C_3. The last carbide was first prepared by Mallett et al.[4] in 1951. We now know that the carbide, first prepared by Moissan, was UC_2, not the sesquicarbide U_2C_3 which was the last to be discovered.

Crystallographic properties

UC is a face-centered cubic compound of the NaCl-type, with a lattice parameter of 4.9605 ± 0.0005 Å as evaluated from numerous measurements made on pure $UC_{1.0}$[5]. Lower values have been reported but these are no doubt due to impurities, particularly nitrogen and oxygen present in solid solution with UC[6]. The X-ray density of UC is 13.63.

UC_2 is always found to be substoichiometric, the composition lying between $UC_{1.85}$ and $UC_{1.94}$[7]. No report exists of a stoichiometric dicarbide, but for simplicity this phase, where the composition is not relevant, will referred to as the UC_2 phase. This phase exists in two crystallographic modifications.

Below 1800 °C the α-structure is tetragonal (space group 14/mm). The scatter in the lattice parameters of UC_2 reported[6] is caused by variations in the heat treament, resulting in the materials having different carbon contents. In fact, since this phase dissolves little oxygen or nitrogen, the lattice parameter is a good measure of the stoichiometry.

The composition $UC_{1.89}$ has $a = 3.519$ Å and $c = 5.979$ Å and the composition $UC_{1.94}$, $a = 3.524$ Å and $c = 5.996$ Å[8]. For the X-ray density of the UC_2 phase the value 11.7 may be taken.

Above 1800 °C UC_2 is transformed into a cubic structure. This, by means of high-temperature neutron diffraction[9], was shown to be of NaCl-type with $a = 5.488$ Å at 1900 °C. The NaCl-type structure of the cubic UC_2-phase is consistent with the observed complete solid solubility between UC and UC_2 at high temperatures (see the phase diagram, p. 170).

The carbide U_2C_3 is body-centered cubic with $a = 8.088 \pm 0.001$ Å; its X-ray density is 12.88[4,10].

Preparation of the uranium carbides

The preparation of UC can be carried out in various ways:

(a) By direct reaction of the elements during arc melting, followed by crushing the cast material to produce a fine powder for sintering. This is not a very economic process and it is difficult to control the reaction to ensure the production of UC without the pick-up of more carbon. However, satisfactory results have been obtained with tungsten electrodes[11].

(b) By reaction of uranium with a hydrocarbon gas (usually methane). This is an attractive approach since the reaction temperature (600–800 °C) is low and the fine powder obtained is in a reactive condition for the sintering. The procedure is to produce uranium hydride, UH_3, by treating bulk uranium with hydrogen at about 250 °C. The hydride is then decomposed above 450 °C to provide a finely divided uranium powder that easily reacts with methane or propane. Methane gives mainly UC at 650 °C and UC_2 above 950 °C (see p. 176).

TABLE 14

PROPERTIES OF URANIUM CARBIDES

Compound	Melting point	Crystal structure	Density
UC	2525 °C	cubic (NaCl-type) $a = 4.9605 \text{ Å}$	13.63
U_2C_3	dec. > 1800 °C	cubic, b.c.c. $a = 8.088 \text{ Å}$	12.88
UC_2	~ 2480 °C	tetragonal (14/mm) $a = 3.524 \text{ Å},$ $c = 5.996 \text{ Å}$ > 1800 °C: cubic (NaCl-type) $a = 5.488 \text{ Å}$	11.7

(c) An economically attractive method for the preparation of UC is the reaction between UO_2 and carbon, according to the overall reaction:

$$UO_2 + 3\,C \rightarrow UC + 3\,CO$$

The reaction proceeds smoothly provided the large volume of CO produced is rapidly removed, either by maintaining a vacuum or a sufficient stream of argon. The CO-pressure of the reaction is given by[12]:

$$\log p_{CO}(\text{atm}) = \frac{-17,000}{T} + 7.37$$

The general procedure is to blend the UO_2 and carbon to give a homogeneous mixture of the correct composition to form UC. The powder is then compacted and heated at $1700\,^\circ\text{C}$ ($p_{CO} = 0.5$ atm) for two hours in a vacuum induction furnace. The pellets are either ground to a fine powder for solid compaction or melted in an arc furnace. It is evident that this method inevitably leads to the presence of some oxygen.

Although U_2C_3 is the stable phase in contact with graphite below about $1500\,^\circ\text{C}$, it cannot be prepared directly, but only by stressing a two phase mixture of $UC + UC_2$ at temperatures between 1300 and $1800\,^\circ\text{C}$ (see p. 171). According to Henney et al.[13] UC_2 is stabilized by dissolved oxygen; when this is removed a transformation of "UC_2" to $U_2C_3 + C$ takes place. A low oxygen potential should therefore be sufficient to ensure the formation of U_2C_3.

Fabrication process

The commercial development of stoichiometric UC has been frustrated by the fact that below $1200\,^\circ\text{C}$ the homogeneity range of UC has a width of less than 0.05 wt. % C. It is difficult to control or adjust the carbon content within such a small limit. This control is necessary because hypostochiometric uranium monocarbide, UC_{1-x}, has an inferior irradiation stability to stoichiometric UC whereas hyperstochiometric uranium monocarbide UC_{1+x}, causes carburization of the cladding. Alloying with, for instance, molybdenum is used to remove either free uranium or an excess of carbon.

The powders produced by any of the proceeding methods are highly reactive to air and moisture and must be stored in an inert atmosphere. For use in nuclear reactors as fuel elements the carbides are fabricated into pellets or rods of suitable size. The densification of these shapes by sintering generally needs a high temperature ($\sim 1800\,^\circ\text{C}$), although it has been shown[14] that high density pellets (95 % theoretical) can be prepared at more convenient temperatures (1500–$1600\,^\circ\text{C}$) provided attention has been paid to particle size and especially to keeping the surface of the fine carbide powders clean during handling. A high-purity, inert atmosphere containing less than 10 ppm oxygen and 10 ppm water vapour is necessary.

Very high densities are of course obtained by melting and casting (98–99% theoretical). The heating is effected by means of an arc in which graphite electrodes are employed. The feed material may be either pre-reacted (cold compacted) UC powder or a mixture of UO_2/C, in which case the reaction takes place in the arc.

Other methods used in the production of UC are hot pressing in graphite dies at 1500 °C, yielding pellets of 98 % theoretical density, isostatic pressing of UC in stainless steel cans, and vibratory compacting[15].

At Eldorado UC is made by the carbothermic reduction of uranium oxide, followed by vacuum arc melting and casting. Vacuum cast rods of UC are available in the "as cast" condition or precisely dimensioned.

The fluidized-bed technique appears to be very attractive for the technological preparation of carbide fuels. The product is a powder which can be converted into pellets or subjected to vibratory compaction. Two processes have been developed: (a) the conversion of oxides by heating at 1600–1800 °C with admixed carbon, (b) the carburization of hydrided metal at 750–850 °C in a methane–hydrogen mixture. The p_{CH_4}/p_{H_2} ratio is important in order to avoid the formation of higher carbides; it should be about 10^{-2} at $1100 °K$[16].

Phase diagram

The phase diagram of the uranium–carbon system has been frequently investigated. The precise fixing of the boundaries of the phases UC, α-UC_2 (tetr.), β-UC_2 (cubic) and U_2C_3 in the system is difficult because conditions of extreme purity are required, since it is now clear that the almost unavoidable presence of oxygen and nitrogen has a significant effect on equilibria in the system.

The phase diagram is shown in Fig. 32. The solubility of carbon in uranium is negligible below 2000 °C. At higher temperatures some carbon is dissolved into the UC lattice, about 185 ppm at the eutectic temperature[17]. The monocarbide UC is stable from room temperature up to its congruent melting point of $2560\pm50\,°C$. Unlike the other cubic carbides, UC has a homogeneity range which was formerly thought to be negligible. At high temperatures a carbon-deficient monocarbide phase UC_{1-x} is formed, causing the lattice to decrease (at 44 % C: 4.949 Å). The solidus is retrograde with a lower limit for the composition $UC_{0.94}$ at about $2000\,°C$.

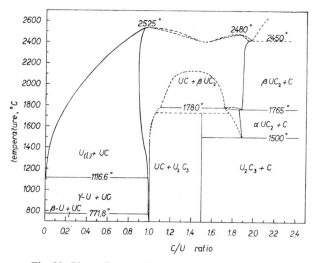

Fig. 32. Phase diagram of the uranium–carbon system.

The stable phase in contact with graphite below $1500\,°C$ is U_2C_3, although it cannot be made directly, but only by heating a two phase mixture of $UC+UC_2$ at temperatures between 1300 and $1800\,°C$. At about $1800\,°C$, U_2C_3 decomposes to $UC+UC_2$ without melting.

UC_2 is the stable phase from $1500\,°C$ to its melting point; it is transformed from the tetragonal α-form to the cubic (NaCl-type) β-form at $1800\,°C$. The transition is so rapid that the cubic phase cannot be quenched in; however its existence has recently been confirmed by a high-temperature neutron diffraction study[9]. At temperatures above $2000\,°C$ complete mutual solubility of UC and UC_2 has been observed which is consistent with the structural resemblance between the two phases.

UC$_2$ is never obtained with the stoichiometric composition, the C/U ratio varying from 1.85 to 1.94. The composition of the UC$_2$–C phase boundary as a function of the temperature was determined by Storms[18], who found the C/U ratio at the phase transition tetragonal $\rightleftarrows$ cubic to be 1.89 and the lower limit at the eutectic temperature (2450 °C) to be 1.94. Under equilibrium conditions UC$_2$ should decompose into U$_2$C$_3$ + C below 1500 °C. However, the UC$_2$ phase is stabilized by dissolved oxygen and only when this is removed, does the transformation 2 UC$_2$ $\rightarrow$ U$_2$C$_3$ + C occur.

Thermodynamic properties

The heat of formation of UC was determined by Storms and Huber[19] from a critical evaluation of previous measurements. They showed that the narrow, but significant, composition range of the monocarbide phase, that is its carbon content, has a marked influence on its heat and free energy of formation. For the composition UC$_{1.0}$ they found $-\Delta H_{298} = 23.2 \pm 0.5$ kcal/mole. But for the composition UC$_{1.90}$ the value $-\Delta H_{298} = 19.2$ kcal/mole, found by these authors, is much lower than the value (21.1 $\pm$ 1.4 kcal/mole) obtained from combustion-calorimetric measurements[20] and the value (20.2 kcal/mole) arrived at in a recent evaluation by Godfrey and Leitnaker[21]. The differences are probably due to uncertainties in the thermal functions of UC$_2$ since the different evaluations made for UC[19,21] appear to agree with one another. The value 19.7 $\pm$ 1.0 kcal/mole is suggested as most

TABLE 15

ENTROPIES OF URANIUM CARBIDES

| | S^0_{298} (cal/deg·mole) | |
	Andon et al.[23]	Westrum et al.[24]
UC	14.03	14.19*
U$_2$C$_3$	32.91	32.95
UC$_{1.91}$	16.31	16.18

*Corrected values[29].

probable. For the heat of formation of U_2C_3 the value $-\Delta H_{298} =$ 50.7 was found[22]; it is a value with a large uncertainty.

Entropies of the uranium carbides, obtained from low-temperature heat capacity measurements[23,24] are in fair agreement (Table 15).

High-temperature heat capacity measurements on UC, made in different laboratories, agree well[25]. For the enthalpy function of $UC_{1.0}$ it was found[5]:

$$H_T - H_{298} = 14.315\,T - 1.5130 \cdot 10^{-4}\,T^2 + 3.5038 \cdot 10^{-7}\,T^3 + \\ + 2.0828 \cdot 10^5\,T^{-1} - 4962.4 \\ (298\text{--}2800\,°K)$$

The heat capacity of α-UC_2 was measured by MacLeod and Hopkins[26] in good agreement with the low-temperature measurements by Andon et al.[23]. The enthalpy function found is:

$$H_T - H_{298} = 17.24\,T + 1.3714 \cdot 10^{-3}\,T^2 + 3.1605 \cdot 10^5\,T^{-1} - 6322 \\ (200\text{--}1500\,°K)$$

The high-temperature thermodynamic properties of uranium dicarbide have been reviewed recently by MacLeod[26].

The enthalpy function found for β-UC_2 is[18]:

$$H_T - H_{298} = -18.73 + 2.944 \cdot 10^{-2}\,T \\ (2050\text{--}2800\,°K)$$

The heat of transition $\alpha \rightarrow \beta$-UC_2 proves from the enthalpy difference at the transition point to be 2.52 kcal/mole.

A number of vaporization studies have been made in the UC phase region. The agreement in the total pressures of these Knudsen experiments is only fair: too high pressures are easily obtained, owing to CO and $UO_{(g)}$ due to contamination with oxygen. Leitnaker and Witteman[7] first showed that UC_2 loses carbon preferentially; a congruent vaporization point was established at $UC_{1.07 \pm 0.05}$. Below 2300 °K, however, it is near $UC_{1.84}$[18]. The uranium containing species in the vapour over UC_2 is now known, from a mass spectrometric study by Norman and

Winchell[27], to consist of both $U_{(g)}$ and $UC_{2(g)}$. The contribution of $UC_{2(g)}$ to the uranium transport, although small, was found to become increasingly important at higher temperatures. This was confirmed by Storms[18] in a comprehensive mass spectrometric study of the U–C system, from which he found:

$$\log p_U(\text{atm}) = \frac{-25,540}{T} + 5.890$$

The free energy of formation of UC, according to the reaction:

$$U_{(l)} + C_{(gr)} \rightarrow UC_{(s)}$$

can be represented by[28]:

$$\Delta G_T^0 = 3.25\,T - 24,600 \qquad (1500\text{–}2000\,^\circ K)$$

For UC_2 the situation is more complicated owing to lack of stoichiometry, the presence of impurities, in particular oxygen and the uncertainties in the thermal functions[29]. Nevertheless for $UC_{1.93}$ the free energy of formation may be represented by:

$$\Delta G_T^0 = -21,500 - 2.8\,T \qquad (1000\text{–}2500\,^\circ K)$$

For U_2C_3 the free energy of formation:

$$\Delta G_T^0 = -48,500 - 2.4\,T \qquad (298\text{–}1800\,^\circ K)$$

may be used[29].

The ternary systems U–C–N *and* U–C–O

During their preparation and in use, the carbides of uranium will be in contact with atmospheres of different compositions. The carbon atoms in the lattice may be replaced by either nitrogen or oxygen. The ease with which dissolved oxygen and nitrogen form solid solutions with the carbides may present a number of problems since it had already been found that these gases seriously affect the properties of the carbides.

(a) *Nitrogen.* A continuous series of solid solutions is formed between UC and UN[30,31]. The replacement of C by N in UC results in a contraction of the lattice according to Vegard's law. The nitrogen pressures over the solid solution $UC_{1-x}N_x$, being lower than over pure UN, have been calculated by Rand[32] assuming the solid solution to be ideal.

Rand has also calculated the nitrogen pressure at which the solid solution $UC_{1-x}N_x$ will react with nitrogen to give a solution of higher nitrogen content:

$$[UC]_{UN} + \tfrac{1}{2}N_2 \rightarrow [UN]_{UC} + C$$

From Table 16 it is seen that, at about 1750 °K, essentially pure UN is formed, starting from pure UC and a nitrogen pressure of 1 atm.

TABLE 16

NITROGEN PRESSURE OVER $UC_{1-x}N_x$ AT 1750 °K[32]

x	p_{N_2} (atm)
0.1	$2.5 \cdot 10^{-5}$
0.5	$2 \cdot 10^{-3}$
0.9	$1.5 \cdot 10^{-1}$

The preparation of the solid solution has been examined[33] and it was found that for practical applications only solid solutions with $x = 0.1$–0.6 can be made easily, when starting from $UO_2 + C$. With increasing amounts of nitrogen it becomes more difficult to obtain homogeneous products.

(b) *Oxygen*. Oxygen is dissolved by UC to a considerable extent. This is of great technological importance, since UC is prepared by reduction of UO_2 according to the reaction: $UO_2 + C \rightarrow UC + CO$. As UO_2 is insoluble in UC, the amount of dissolved oxygen will depend largely on the rate at which the UO_2 can be converted into the easily soluble, but unstable, UO which goes into solid solution with UC. The reaction between UO_2 and UC proceeds smoothly above 1600 °C with removal of CO. The CO pressure over U(C, O) compositions was[12]:

$$\log p_{CO}(\text{atm}) = \frac{-17,000}{T} + 7.37$$

The lattice parameter of the $UC_{1-x}O_x$ phase passes through a maximum at $x = 0.027$. The limit of oxygen solubility appeared when $x = 0.37$ at $1200°C^{34}$.

Oxygen is also soluble in the tetragonal UC_2 structure, where it stabilizes this phase at temperatures below $1500°C^{13}$, conditions in which U_2C_3 is the stable phase. It was found that annealing a specimen of U_2C_3 in a poor vacuum (10^{-2} mm) at $1500°C$ for 20 hours resulted in the disappearance of U_2C_3 and the formation of the "UC_2"-phase. A low oxygen potential should therefore be sufficient to ensure the formation of U_2C_3. The difference in the free energy of formation of oxygen-saturated UC_2 and U_2C_3 must be small and has been estimated to be about 5 kcal/mole[35]. This was confirmed in a recent determination[36] yielding at $1000°K$ for pure UC_2, $\Delta G^0 = -23.6$ kcal/mole, and for oxygen stabilized "UC_2", $\Delta G^0 = -29.0$ kcal/mole. The small difference in the free energies of these carbide phases might explain why the application of stress initiates the transformation:

$$\text{"}UC_2\text{"} \rightarrow U_2C_3 + C$$

The mechanism of the formation of U_2C_3 has been examined in detail by Nickel and Saeger[37].

Chemical properties of the uranium carbides

The reactions of the uranium carbides with metallic and non-metallic elements are of considerable technological interest.

When considering these carbides as a nuclear fuel, it is of primary importance to know their behaviour when exposed to gases such as nitrogen, oxygen and reactor coolants. A knowledge of their compatibility with metals is of special interest in the development of inert cladding materials.

(*a*) *Oxygen.* At low temperatures ($< 150°C$) and low partial pressures, UC is oxidized to UO_{2+x} and C, whereas at higher temperatures partial oxidation of the carbon occurs, which is complete at about $400°C$. The reaction: $3\,UC + 7\,O_2 \rightarrow U_3O_8 + 3\,CO_2$ is highly exothermic (358 kcal/mole of UC). The ignition of UC has been studied as a function of temperature and particle properties[38]. It was shown by Dell and Wheeler that the final product of the oxidation below 330 °C is UO_3 rather than UO_2 or U_3O_8[39]. Some carbon is retained in the lattice in the form of CO_2, the composition being $UO_3 \cdot 0.3CO_2$. An infrared examination of the oxide product showed absorption peaks characteristic of CO_2 chemisorbed on the oxide. On heating the sample in vacuum CO_2 was evolved.

(b) *Hydrogen.* The equilibrium:

$$UC_2 + 2 H_2 \rightleftarrows UC + CH_4$$

is of importance in the fluidized-bed preparation of UC. Thus, at 1100 °K the ratio p_{CH_4}/p_{H_2} should not be above $5 \cdot 10^{-3}$, then no deposition of carbon takes place[40].

(c) *Water.* Uranium monocarbide does not react with water below 50–60 °C because of the formation of a protective layer. Above 60 °C a rapid reaction takes place of which the products are mainly UO_2 and CH_4, together with a large number of higher hydrocarbons. UC_2 behaves in a similar way, the predominant product here being C_2H_6 (Table 17).

TABLE 17

VOLATILE PRODUCTS (IN VOL. %) FROM THE HYDROLYSIS OF UC
AND UC_2 WITH WATER AT 100 °C[41]

	H_2	CH_4	C_2H_4	C_2H_6	C_3H_8	nC_4H_{10}
UC	12.1	84.9	0.25	2.0	0.6	0.15
UC_2	25.4	19.0	4.4	38.6	2.1	4.8

During hydrolysis of UC the free radicals $=CH_2$ and $-CH_3$ are formed[41] together with atomic hydrogen. The last can hydrogenate a free radical or combine to molecular hydrogen. In UC_2 the distance between the carbon atoms (1.34 Å) is close to the distance of the $C=C$ bond, thus yielding C_2H_4 and C_2H_6 from the free radicals $=C=C=$.

(d) *Nitrogen.* UC and UN form a continuous series of solid solutions the composition depending on the nitrogen pressure and the temperature. Contamination with nitrogen may result during arc melting when nitrogen is present in the gas. It was shown by Rand[32] that a nitrogen pressure of about 1 mm at 1500 °C will produce a solid solution $UC_{1-x}N_x$ with $x = 0.5$.

Quite small amounts of nitrogen can stabilize UC against formation of a higher carbide in the presence of graphite.

The reaction of uranium monocarbide with nitrogen below 1050 °C has been investigated by Hanson[42]. The reaction rate varies with the $\frac{3}{4}$ power of the nitrogen pressure. The activation energy of the reaction

$$UC + \tfrac{3}{4} N_2 \rightarrow \tfrac{1}{2} U_2N_3 + C$$

is 31.8 ± 1.3 kcal/mole.

(*e*) *Compatibility with metals.* The reaction between carbides and metals presents a serious problem when considering metals as possible cladding materials. Because of the metallic nature of the carbides, such interaction is extensive. This is a drawback of the carbides of uranium since it limits the selection of the cladding materials. Many systems have been investigated and the information obtained, in addition with thermodynamic data, has been used to predict the compatibility between UC and various refractory metals[29].

Nickel and its alloys, aluminium and beryllium remain compatible with the monocarbide only below 500 °C (a temperature appropriate to organic-cooled reactors). Iron, chromium and stainless steel exhibit no significant reaction up to 900 °C. It should be noted, however, that hyperstoichiometric UC becomes incompatible with these metals in the presence of sodium above 600 °C (for instance, in sodium-cooled, graphite-moderated reactors). The carburization takes place at rates determined by the temperature and the amount of UC_2 present in the monocarbide. Carbon transfer occurs *via* sodium to the stainless steel and the UC_2 phase, dispersed in the UC matrix, is reduced to UC[43]. Stoichiometric monocarbide itself is stable to sodium up to at least 900 °C, thus the latter is an attractive coolant, for instance in fast reactors with (U, Pu)C fuels.

References p. 182

Niobium and its alloys are compatible with UC up to high temperatures and thus are of interest for high-temperature, sodium-cooled power reactors. At 1800 °C niobium reduces UC to uranium and NbC.

It is interesting to note that the chemical reactivity of the carbides, a drawback in other respects, may also be an advantage. For example, alloying with molybdenum, chromium or vanadium (1–3%) is used to remove either free uranium or excess of carbon in the monocarbide, because, as already said, stoichiometric monocarbide has optimum properties. Moreover, the solubility of metals such as zirconium, niobium, titanium and vanadium in UC leads to enhanced mechanical properties and to better corrosion resistance. Thus, UC alloyed with 50 mole % ZrC shows a markedly improved resistance to attack by water, steam and air. Furthermore, alloying with ZrC raises the melting point from 2390 °C to about 3000 °C. The (U, Th)C$_2$ solid solution has found application, when dispersed in graphite, as a fuel in the Dragon reactor.

Physical properties

Uranium monocarbide exhibits typical metallic behaviour in its thermophysical properties. The thermal conductivity of dense UC goes through a minimum at about 300 °C, after which it has a slight positive temperature coefficient; 0.055 cal/cm.sec.deg. may then be used as a mean value up to 1200 °C. Recently, Leary et al.[44] showed the thermal conductivity to increase linearly in the temperature region from 300–460 °C and found:

$$k = 0.0512 + 7.3 \cdot 10^{-6} t$$

(k in cal/cm.sec.deg. and t in °C), in good agreement with previous findings[45,46]. However, measurements of the thermal conductivity of UC shows a large scatter due to variations in the carbon content. According to Moser and Kruger[47] the conductivity drops from 0.057 cal/cm.sec.deg. at room temperature to 0.041 cal/cm.sec.deg. at 400 °C and then rises gently to reach a value of 0.042 cal/cm.sec.deg. at 700 °C.

The thermal conductivity of UC is about five times higher than the value for UO_2. This is of much importance in the application of UC in high-temperature reactors, although it should be noted that replacement of uranium by plutonium causes the thermal conductivity to decrease[44]. However, at high temperatures the thermal conductivity of the solid, for example one with 20 mole % PuC, should approach that of UC since the temperature coefficient of the former is much larger than that of the latter.

The electrical resistivity of UC, as with metallic conductors, increases with increasing temperature[44]. Grossman[48] found:

$$\sigma = 20.4 \cdot 10^{-6} + 114.8 \cdot 10^{-9} \, T \, \Omega \text{cm} \quad (1200-2050\,°\text{K})$$

The thermal expansion coefficient of UC is $10.5 \cdot 10^{-6}$ per °C.

The carbides UC and UC_2, although they are metallic conductors, are easily hydrolized by water and dilute acids, giving a variety of hydrocarbons. In UC each uranium atom is surrounded by an octahedral disposition of carbon atoms. The simplest picture of the structure is that of an ionic lattice, containing U^{4+} ions and C^{4-} ions; any excess of electrons constitutes a metallic-like bond. However, arguments from ionic and covalent radii and the absence of a magnetic anomaly, suggest a model with a large amount of covalent bonding. On this view, the undoubted existence of a conduction band may be explained by an overlap of the 7s and 5f orbitals.

Uranium silicides

Recently, the uranium silicides have become of much interest in nuclear technology. This is so particularly for U_3Si which combines a high density (15.6) with a low parasitic neutron absorption and a good aqueous-corrosion resistance[49]. Preliminary irradiation tests have shown it to be a useful potential nuclear fuel.

Phase diagram

In the uranium–silicon system, six phases have been identified[50]: U_3Si, U_3Si_2, USi, α-USi_2, β-USi_2 and USi_3. However, the composition of the β-USi_2 phase is in doubt and the formula

References p. 182

U_2Si_3 has been suggested instead[51]. The relevant crystallographic properties of these phases are listed in Table 18.

TABLE 18

CRYSTALLOGRAPHIC PROPERTIES OF URANIUM SILICIDE PHASES

Phase	Structure	Lattice parameter (Å)	Density	Melting point
U_3Si	tetragonal (I4/mcm)	$a = 6.030$ $c = 8.696$	15.58	transf. (765 °C)
U_3Si	cubic	$a = 4.346$		930 °C, dec.
U_3Si_2	tetragonal (P4/mbm)	$a = 7.330$ $c = 3.900$	12.2	1665 °C
USi	o. rhombic	$a = 5.66$ $b = 7.66$ $c = 3.91$	10.4	1575 °C, dec.
U_2Si_3 (or β-USi_2)	hexagonal (C6/mmm)	$a = 3.85$ $b = 4.06$	9.25	1610 °C, dec.
α-USi_2	tetragonal (I4/amd)	$a = 3.97$ $c = 13.71$	8.98	1700 °C
USi_3	cubic	$a = 4.03$	8.15	1510 °C, dec.

Phase relationships in the uranium–silicon system are shown in Fig. 33. The solubility of silicon in α-U is negligible; the transformation temperature $\alpha \rightarrow \beta$ uranium remains unchanged. The $\beta \rightarrow \gamma$ transformation is raised by dissolved silicon (< 1 at. %) in β-U, from 770° to 795 °C[51]. The maximum solubility of silicon in γ-U is about 1.75 at. % at 980 °C. The silicide U_3Si decomposes at 930 °C into γ-U and U_3Si_2, and the two form a eutectic at 985 °C.

The preparation of U_3Si

The existence of U_3Si, often called the δ-phase, which is much denser than the other phases in the system, was not at first suspected. This silicide is not formed except by heating uranium and U_3Si_2 at about 800 °C, according to the reaction:

$$3 U + U_3Si_2 \rightarrow 2 U_3Si$$

The reaction is slow and homogeneity is difficult to obtain. Therefore, annealing at 800 °C ("deltising") is necessary to complete the transformation.

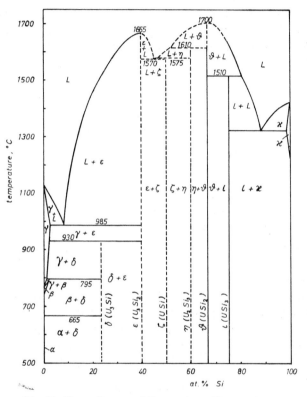

Fig. 33. Phase diagram of the uranium–silicon system.

Rods of the U_3Si are produced at Eldorado (Canada) by the vacuum induction melting of uranium and silicon, followed by casting into cored rods. The rods are then heat treated to transform the as-cast structure of U_3Si_2 in uranium into the corrosion resistant U_3Si phase[52]. U_3Si has a narrow composition range, thus a single phase alloy can rarely be obtained in practice, excess of uranium or U_3Si_2 often being present.

References p. 182

Properties of U_3Si

The thermal conductivity of U_3Si is 0.15 W/cm.sec.deg. at 25 °C. Over the temperature range of 0 to 1000 °C, the value 0.2 W/cm.sec.deg. may be taken as the average value. The thermodynamic properties of U_3Si and of the other silicide phases have been reviewed recently[53]. The heats of formation of the uranium silicides were measured by Gross *et al.*[54]; the results are listed in Table 19.

TABLE 19

HEATS OF FORMATION OF URANIUM SILICIDES[54]

	$-\Delta H_{298}$ (*kcal/mole*)
U_3Si	—
U_3Si_2	40.7
USi	19.2
USi_2	31.1
USi_3	31.6

It was found by Blum *et al.*[55] that the tetragonal U_3Si phase undergoes a phase transition to a cubic phase at 765 °C. The lattice parameter of this phase is 4.346 Å at 780 °C.

Up to the present, the major obstacle to the use of U_3Si as a nuclear fuel has been its swelling under irradiation which is possibly due to a disordering of the tetragonal lattice. Recently, however, promising results for irradiation tests have been reported[49].

REFERENCES

1   H. MOISSAN, *Ann. Chim. Phys.*, 9 [7] (1896) 302.
2   L. M. LITZ, A. B. GARRETT AND F. C. CROXTON, *J. Am. Chem. Soc.*, 70 (1948) 1718.
3   R. E. RUNDLE, N. C. BAENZIGER, A. S. WILSON AND R. A. M. McDONALD, *J. Am. Chem. Soc.*, 70 (1948) 99.
4   H. W. MALLET, A. F. GERDS AND D. S. VAUGHAN, *J. Electrochem. Soc.*, 98 (1951) 505.
5   E. K. STORMS, *The Refractory Carbides*, Academic Press, New York and London (1967), p. 171.
6   *The Uranium–Carbon and Plutonium–Carbon Systems*, IAEA, Technical Reports Series, No. 14. Vienna (1963).

7   J. M. LEITNAKER AND W. G. WITTEMAN, *J. Chem. Phys.*, 36 (1962) 1445.
8   W. G. WITTEMAN AND A. L. BOWMAN, *TID-Report* 7676 (1963).
9   A. L. BOWMAN, G. P. ARNOLD, W. G. WITTEMAN, T. C. WALLACE AND N. G. NERESON, *Acta Cryst.*, 21 (1966) 670.
10 A. E. AUSTIN, *Acta Cryst.*, 12 (1959) 159.
11 *LA-Report* 2942 (1964), p. 187.
12 J. BAZIN AND A. ACCARY, *Proc. Brit. Ceram. Soc.*, 8 (1967) 175.
13 J. HENNEY, N. A. HILL AND D. T. LIVEY, *Trans. Brit. Ceram. Soc.*, 62 (1963) 955.
14 J. R. McLAREN, M. C. REGAN AND H. J. HEDGER, *Carbides in Nuclear Energy*, Macmillan and Co., London, Vol. II (1964), p. 588.
15 A. ACCARY, *J. Nucl. Mat.*, 8 (1963) 281.
16 *ANL-Report* 7175 (1966), p. 127.
17 W. B. WILSON, *J. Am. Ceram. Soc.*, 43 (1961) 76.
18 E. K. STORMS, *Thermodynamics, Proceedings of a Symposium (1965)*, IAEA, Vienna, Vol I, (1966), p. 309.
19 E. K. STORMS AND E. J. HUBER, JR., *J. Nucl. Mat.*, 23 (1967) 19.
20 E. J. HUBER, Jr., E. L. HEAD AND C. E. HOLLEY, JR., *J. Phys. Chem.*, 67 (1963) 1730.
21 J. M. LEITNAKER AND T. G. GODFREY, *J. Chem. Eng. Data*, 11 (1966) 392.
22 W. K. BEHL AND J. J. EGAN, *J. Electrochem. Soc.*, 113 (1966) 376.
23 R. J. L. ANDON, J. F. COUNSELL, J. F. MARTIN AND H. J. HEDGER, *Trans. Faraday Soc.*, 60 (1964) 1030.
24 E. F. WESTRUM, JR., E. SEUTS AND H. K. LONSDALE, *Advances in Thermophysical Properties at Extreme Temperatures and Pressures*, American Society of Mechanical Engineers, New York (1965), p. 156.
25 J. M. LEITNAKER AND T. G. GODFREY, *J. Nucl. Mat.*, 21 (1967) 175.
26 A. C. MacLEOD AND S. W. J. HOPKINS, *Proc. Brit. Ceram. Soc.*, 8 (1967), p. 15; see also A. C. MacLEOD, *J. Inorg. Nucl. Chem.*, 31 (1969) 715.
27 J. H. NORMAN AND P. WINCHELL, *J. Phys. Chem.*, 68 (1964) 3802.
28 J. BAZIN AND A. ACCARY, *Proc. Brit. Ceram. Soc.*, 8 (1967) 175.
29 CH. E. HOLLEY, JR. AND E. K. STORMS, *Thermodynamics of Nuclear Materials, Proceedings of a Symposium (1967)*, IAEA, Vienna (1968), p. 397.
30 J. WILLIAMS AND R. A. J. SAMBELL, *J. Less-Common Metals*, 1 (1959) 217.
31 A. E. AUSTIN AND A. F. GERDS, *BMI-Report* 1272 (1958).
32 M. H. RAND, *AERE-M-Report* 1360 (1964).
33 A. NAOUMIDIS AND H. J. STÖCKER, *Thermodynamics of Nuclear Materials, Proceedings of a Symposium (1967)*, IAEA, Vienna (1968), p. 287.
34 P. MAGNIER, J. TROUVÉ AND A. ACCARY, *Carbides in Nuclear Energy*, Macmillan and Co., London, Vol I, (1964) p. 95.
35 *The Uranium–Carbon and Plutonium–Carbon Systems*, IAEA, Vienna, Technical Reports Series, No. 14 (1963), p. 39.
36 E. J. McIVER, *AERE-R-Report* 4983 (1966).
37 H. NICKEL AND H. SAEGER, *J. Nucl. Mat.*, 28 (1968) 93.
38 R. M. DELL AND V. J. WHEELER, *J. Nucl. Mat.*, 21 (1967) 328.
39 R. M. DELL, V. J. WHEELER AND E. J. McIVER, *Trans. Faraday Soc.*, 62 (1966) 3591.

40   R. G. SOWDEN, N. HODGE, M. J. MORETON-SMITH AND D. B. WHITE, *Carbides in Nuclear Energy*, Macmillan and Co., London, Vol. I (1964), p. 297.

41   J. BESSON, P. BLUM AND J. SPITZ, *ibid.*, p. 273.

42   L. A. HANSON, *J. Nucl. Mat.*, 19 (1966) 15.

43   P. E. ELKINS, *NAA-SR-Report* 7502 (1964).

44   J. A. LEARY, R. L. THOMAS, A. E. OGARD AND G. C. WONN, *Carbides in Nuclear Energy*, Macmillan and Co., London, Vol. I (1964), p. 365.

45   A. C. SECREST, JR., E. L. FOSTER AND R. F. DICKERSON, *BMI-Report* 1309 (1959).

46   F. A. ROUGH AND W. CHUBB, *BMI-Report* 1441 (1960).

47   J. B. MOSER AND O. L. KRUGER, *J. Appl. Phys.*, 38 (1967) 3215.

48   L. N. GROSSMAN, *J. Am. Ceram. Soc.*, 46 (1963) 264.

49   *AECL-Report* 2874 (1967).

50   W. H. ZACHARIASEN, *Acta Cryst.*, 2 (1949) 94.

51   A. R. KAUFMAN, B. CULLITY AND G. BITSIANES, *Trans. AIME, J. Metals*, 209 [9] (1957) 23.

52   *AECL-Report* 2684 (1967).

53   M. H. RAND AND O. KUBASCHEWSKI, *The Thermochemical Properties of Uranium Compounds*, Oliver & Boyd, Edinburgh and London (1963), p. 46.

54   P. GROSS, C. HAYMAN AND H. CLAYTON, *Thermodynamics of Nuclear Materials, Proceedings of a Symposium (1962)*, IAEA, Vienna (1963), p. 653.

55   P. L. BLUM, G. SILVESTRE AND H. VAUGOYEAU, *Compt. Rend.*, 260 (1965) 5538.

CHAPTER 11

Compounds of Uranium with Elements of Group V

Introduction

Interest in the properties of the compounds of uranium with the Group V elements, the uranium pnictides, has much increased in the past few years. This is especially the case with the monopnictides UN and UP, which are, because of their good physico-chemical properties and refractory characteristics, potential nuclear fuel materials. The monopnictides all have the NaCl-type structure and the melting points, as given in Table 20, indicate high binding energies, higher than the corresponding compounds from Group IV and VI. This is also indicated both by the hardness and the heats of formation, although complete thermodynamic information is still lacking.

TABLE 20

PHYSICAL PROPERTIES OF THE URANIUM MONOPNICTIDES

Compound	Lattice parameter ($\mathring{A}$) (cubic, NaCl-type)	Density	Melting point, $°C$
UN	4.889	14.32	2850° (2.5 atm)
UP	5.589	10.27	2610°
UAs	5.779	10.81	2450°
USb	6.2091	9.98	1850°
UBi	6.364?		1250°

The monopnictides have high electrical conductivity, they are of metallic nature, and the chemical bonding in these semimetallic compounds presents an interesting problem, since their consideration on a completely ionic basis is not valid[1]. In uranium mononitride a type of mixed bonding is assumed, the mean uranium valency being somewhere between 4 and 6.

As the N/U ratio increases, a reduction in net magnetic moment is indicated by the decreasing susceptibility, showing an increase in U^{6+} ions. Indeed extrapolation of magnetic moment suggests that UN_2 should be diamagnetic.

The uranium monopnictides all show an antiferromagnetic ordering of the unpaired spins; the Néel temperature increases with increasing atomic number (Table 21).

TABLE 21

NÉEL TEMPERATURE OF THE URANIUM MONOPNICTIDES

UN	UP	UAs	USb	UBi
52°K	123°K	128°K	213°K	—

Apart from these considerations, the combination of a high melting point with good thermal conductivity and good thermal stability, even under neutron irradiation, make these monopnictides, in particular UN, superior to uranium dioxide and uranium monocarbide. This has therefore opened up a prospect of their application in power reactors.

Uranium nitrides

The first uranium nitride was prepared in 1842 by Rammelsberg[2] by the reaction of UCl_4 and ammonia. The nitrogen containing compounds, obtained since, were long assumed to be U_3N_4, U_5N_4 and U_5N_2; indeed, until Rundle et al.[3] characterized them as UN, U_2N_3 and UN_2 in 1948.

Crystallographic properties

The compound UN is face-centered cubic, NaCl-type, with a lattice parameter of 4.889 ± 0.001 Å; its theoretical density is 14.32. U_2N_3 is body-centered cubic (Mn_2O_3-type) with $a =$

10.678 ± 0.005 Å; its X-ray density is 11.24. UN_2 is face-centered cubic (CaF_2-type); with $a = 5.31$ Å; the X-ray density is 11.73.

Although U_2N_3 and UN_2 are crystallographically dissimilar, a solid solution between these compounds has been recognized[3] in which, as the nitrogen content is increased, the U_2N_3 structure is transformed into the UN_2 structure; to attain this only slight changes in the nitrogen positions are required.

A fourth uranium nitride phase was first suggested by Vaughan[4]. The phase is hexagonal (La_2O_3-type) with lattice parameters $a = 3.698$ Å and $c = 5.839$ Å and is isomorphous with Th_2N_3. This polymorphic form of Rundle's U_2N_3 will be indicated here as U_2N_3 (hex.); its existence has since been confirmed[5, 6, 7].

Preparation of the uranium nitrides

Metallic uranium, either in the mass or as a powder, reacts with nitrogen at temperatures above 400 °C to give various nitrides or mixtures of nitrides. Amongst these are the nitrides U_2N_3 and UN_2 which are not stable at high temperatures. Above 1200 °C, only the nitride UN is stable at nitrogen pressures of one atmosphere.

Uranium mononitride, UN, a light-grey powder, may be prepared in a pure form by the decomposition of uranium sesquinitride, U_2N_3, in vacuum or in an inert atmosphere at 1250–1500 °C. The higher the decomposition temperature, the less sensitive to oxygen is the UN powder.

Direct conversion of uranium metal to UN by arc melting in nitrogen (20 atm) has been successful, a high nitrogen pressure is necessary in order to prevent its decomposition. Uranium, in the form of a cylindrical rod, served as the electrode; this was arc melted inside a deep cylindrical water-cooled copper crucible[8].

A technically attractive method is the reaction of UO_2, or ammonium uranate, with nitrogen in the presence of carbon at 1700 °C:

$$UO_2 + \tfrac{1}{2} N_2 + C \xrightarrow{1700\,°C} UN + CO$$

The impurities in the nitride, however, are high compared with those in products from the previous methods. In fact, a solid solution of UO and UC in UN is formed[9].

An interesting method for the preparation of UN is the reaction in which UF_4 as the starting material is employed (see p.198), since in this way enriched UN in a pure form may be obtained.

The densification of UN powders is difficult to achieve by normal sintering techniques, since sintering below 1600°C does not yield compacts of more than 80% of theoretical density and sintering at higher temperatures (> 1700°C) causes decomposition of the mononitride.

To produce dense UN compacts isostatic hot pressing of precompacted UN has been used, only then does sintering of UN lead to high densities when external pressure is applied[10,11]. The pre-compacted UN rods are canned in metal tubes which are then heated to 1500°C for 3 hours under an external pressure. UN bodies with densities as high as 96 % of the theoretical density have been obtained in this way.

Uranium sesquinitride, cubic U_2N_3 may be prepared by the reaction of nitrogen with uranium or by successively hydriding and dehydriding the metal before allowing it to react with nitrogen. The reaction between the elements becomes marked at 500–600°C, yielding the cubic U_2N_3 phase with a composition of about $UN_{1.6}$. Reaction rates have been measured by Mallet and Gerds[12]. With finely divided uranium or with UH_3 reaction occurs even at 200°C and yields products with a higher N/U ratio (1.7–1.8).

The reaction between uranium and nitrogen is strongly exothermic. Therefore, with finely divided uranium (obtained by decomposition of UH_3), conditions must be carefully controlled since otherwise the temperature rises very steeply in a few seconds, causing sintering of the material. It was found[13] that such control can easily be secured when the reaction is carried out in the presence of hydrogen—the partial pressure of H_2 being 0.1–50 mm Hg. The partial pressure of the nitrogen has only a slight effect upon the rate of its reaction with uranium, the reaction rate is determined by the partial pressure of hydrogen. The resulting product has a composition of about $UN_{1.75}$.

The composition of U_2N_3 depends on the reaction conditions, the N/U ratio varying from 1.6 to 1.8. Excess of nitrogen can be removed by heating the sesquinitride in vacuum at 950 °C until the constant pressure of the 3-phase equilibrium $(UN + UN_{1.5+x} + N_2)$ is obtained. At this temperature the N/U ratio is 1.54, according to the phase relationships discussed below.

Hexagonal U_2N_3 has been made by the reaction of UN with nitrogen (0.4 atm) at 1315 °C[7] and also by heating U_2N_3 (cubic) in vacuum $(10^{-2}$ mm) for one hour at 1100 °C[5].

Actually, uranium dinitride, UN_2, has not been obtained in a pure form; evidence for its existence is based entirely on X-ray data. In samples with the composition above N/U = 1.8 the presence of UN_2 has been recognized[3, 14] but, just as with cubic UC_2, there is evidence that the composition of the phase is restricted to $UN_{1.86}$ (see p. 192).

Phase relationships

The solubility of nitrogen in uranium increases from 10 ppm at the melting point of uranium (1132 °C) to 4.6 wt. % at the melting point of UN[15]. The solubility expressed in atom fraction of nitrogen is:

$$\log X_N = \frac{-8480}{T} + 2.4143$$

From room temperature to about 1100 °C the phase UN has a relatively narrow range of stoichiometry. At higher temperatures it takes up considerable amounts of nitrogen or uranium. Benz and Bowman[7] found that UN is very similar to UC in that the lower phase boundary shows retrograde behaviour with an upper limit of $UN_{0.96}$ at 1500 °C and $UN_{0.80}$ at 2000 °C.

The dissociation pressure of UN, according to the equilibrium:

$$UN(s) \rightleftarrows U(s) + \tfrac{1}{2} N_2(g)$$

has been measured by several investigators[15-20]. The results are in good agreement and can be summarized by:

$$\log p_{N_2}(atm) = \frac{-31{,}580}{T} + 9.80$$
$$(1750–2750 °K)$$

The nitrogen pressure as a function of the temperature, plotted as $\log p_{N_2}$ versus $1/T$ (fig. 34) is not linear over the whole temperature range. Above 2500 °K, the solubility of nitrogen in uranium causes a varying uranium activity with varying nitrogen solubility in the liquid uranium. The melting point of UN was found to be 2850 °C at a pressure of 2.5 atm nitrogen[15,16].

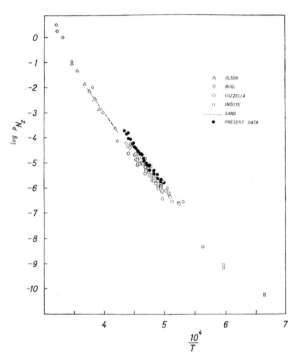

Fig. 34. The dissociation pressure of uranium mononitride (ref. 20).

According to Bugl and Bauer[15] the lower limit of the U_2N_3 phase is $UN_{1.54}$ at 850 °C and $UN_{1.53}$ at 1250 °C. Somewhat higher values (Fig. 35) have been found by other investigators[18,21].

A further point of disagreement in the phase diagram concerns the cubic-hexagonal phase transition in U_2N_3. Benz and Bowman[7] observed a reversible transition at 1120 °C, in good agreement with the observations of Müller et al.[22] who found the transition temperature to be 1085 °C. However, the formation of hexagonal U_2N_3 has also been noted at 650–850 °C[12]. Rand[23] proved that the cubic phase is not reformed when the hexagonal phase is annealed below the transition temperature. Rand's observation has been confirmed by the detailed study of Naoumidis[24], who

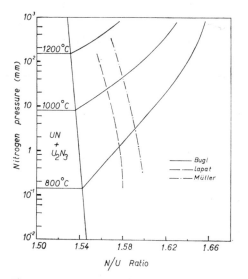

Fig. 35. Phase diagram of the UN–U_2N_3 system.

has shown that the hexagonal (β) phase has a homogeneity range from about U_3N_4 to U_2N_3. This means that the cubic and hexagonal phases may not be considered as polymorphs since their conversion occurs as a result of a change in the composition. Thus, the hexagonal phase may also be stable below 900 °C if low nitrogen pressures are employed.

For the nitrogen pressures of the three-phase equilibrium UN, $UN_{1.5+x}$, N_2 it has been found[22]

$$\log p_{N_2}(\text{atm}) = \frac{-13,000}{T} + 8.16$$

$$(1050\text{–}1358 \,^\circ K)$$

in good agreement with other measurements[15,17,21].

A solid solution of nitrogen in U_2N_3 (cubic) is formed as the nitrogen content of the material is increased. This results in a lattice contraction from 10.688 Å ($UN_{1.50}$) to 10.636 Å ($UN_{1.75}$). The solid solution is reported[3] to be transformed into a fluorite phase (UN_2) above N/U = 1.8.

Indeed, it has been found that material with a composition of $UN_{1.75}$ at 600 °C, when heated under a pressure of 1800 psi, produces a mixture of UN and UN_2[25]. Some recent work, however,

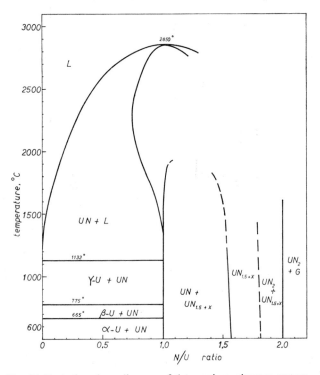

Fig. 36. Tentative phase diagram of the uranium–nitrogen system.

has failed to prepare UN_2 either by treatment of U_2N_3 at high nitrogen pressures (140 atm) or by the action of ammonia on UH_3. The upper limit found was $UN_{1.84}$. It has been suggested therefore[24] that the limiting composition UN_2 cannot be reached and, just as the cubic UC_2 phase, exists up to $UN_{1.86}$ only.

Although the complete phase relationships are not finally settled, the phase diagram shown in Fig. 36 gives an accurate survey of the results at present available.

Thermodynamic properties

The heats of formation of the uranium nitrides have been measured by Gross *et al.*[26], who found for UN that $-\Delta H_{298} = 69.6 \pm 0.4$ kcal/mole. This is in reasonable agreement with the value $-\Delta H_{298} = 72.0 \pm 1.2$ kcal/mole obtained from oxygen-bomb calorimetry and the value $-\Delta H_{298} = 72.8 \pm 1.2$ kcal/mole obtained from fluorine-bomb calorimetry[27]. The value $-\Delta H_{298} = 72.4 \pm 1.0$ has been suggested[27] as the most probable one.

For U_2N_3, the value $-\Delta H_{298} = 169.4$ kcal/mole is derived from the dissociation pressure measurements of U_2N_3 in equilibrium with UN. For the heat of formation of (hypothetical) UN_2 the value $-\Delta H_{298} = 107 \pm 3$ kcal/mole may be derived from the heats of formation of the other nitride phases.

Low-temperature heat-capacity measurements have been made by Westrum *et al.*[28] and by Counsell *et al.*[29]. The two sets of results on UN are in good agreement. The mean value for the entropy of UN follows from these measurements: $S^0_{298} = 14.90$ cal/deg.mole. For U_2N_3 only Counsell's results are available; he found for $UN_{1.59}$ the value $S^0_{298} = 15.54$ cal/deg.mole and for $UN_{1.73}$ $S^0_{298} = 15.74$ cal/deg.mole.

Heat capacity data on UN at high temperatures have been derived from enthalpy measurements in a drop calorimeter[10,20]. The results obtained are:

$$C_p = 13.32 + 1.19 \cdot 10^{-3} T - 2.10 \cdot 10^5 T^{-2} \text{ cal/deg.mole}$$

$$(273–1423\,°K)$$

The value of C_p at 298.15 °K (11.31 cal) is a good agreement with the low temperature heat capacity measurements.

The thermodynamic properties of UN, as given in Table 22, have been calculated[20] from the data given above.

TABLE 22

THERMODYNAMIC PROPERTIES OF UN[20]

°K	$H_T - H_{298}$ (cal/mole)	S_T^0 (cal/deg.mole)	$-(G^0 - H_{298})/T$ (cal/deg.mole)
400	1204	18.44	15.43
500	2450	21.22	16.32
600	3743	23.57	17.33
700	5077	25.63	18.38
800	6448	27.46	19.40
900	7855	29.12	20.39
1000	9295	30.63	21.34
1100	10770	32.04	22.25
1200	12280	33.35	23.12
1300	13817	34.58	23.95
1400	15390	35.75	24.76
1500	16994	36.86	25.53

$S_{298}^0 = 14.97$ cal/deg.mole.

The solubility of nitrogen in U_2N_3 has also been studied[15,21,22]. In contrast with the results for the two-phase region, the agreement between these measurements is only fair.

Magnetic properties and chemical bonding

Magnetic measurements on material in the composition range $UN_{1.5}$ to $UN_{1.86}$[30,31] show that the susceptibility falls with increasing nitrogen content; this indicates a reduction in the net magnetic moment. The temperature of magnetic ordering falls from 94 °K ($UN_{1.59}$) to 33 °K ($UN_{1.73}$) and the heat and entropy of the transition become smaller[27]. Extrapolation shows that UN_2 should be diamagnetic, suggesting the presence of U^{6+} ions. In agreement with this is the finding that the dissolution of nitrogen in U_2N_3 causes a lattice contraction from 10.688 Å ($UN_{1.50}$) to 10.636 Å ($UN_{1.75}$), and also the small lattice constant of UN_2 ($a = 5.21$ Å) compared with that of UO_2 ($a = 5.46$ Å). Apparently, a gradual transition of U^{4+}, or U^{5+} ions, to U^{6+} ions takes place as the nitrogen content is increased. It is interesting to note the thermal anomaly in UN at 52 °K which is due to an antiferromagnetic transition. This is significantly different from that in UC, where a magnetic transition has not been found. Thus, a change in bond character from UC through UN and UP to US has been predicted[30]; in this series UC has a covalent and US an ionic character, whereas UN has considerable covalency.

Physical properties of UN

The thermal conductivity of uranium mononitride has been measured by Keller[32], who found

$k = 0.0294 + 0.306 \cdot 10^{-4} t$ (k in cal/cm.sec.deg, t in °C).

The thermal conductivity thus rises from 0.030 at room temperature to 0.060 at 1000°C. Although UN has a good thermal conductivity, it is a relatively poor electrical conductor, having resistivity values about four times greater than those of uranium metal. For the electrical resistivity of UN between room temperature and 1000°K the mean value of 200 μ Ωcm may be taken. It is reported to be a p-type conductor[14]. The thermal expansion coefficient of UN is $8.6 \cdot 10^{-6}$ per °C.

The rate of self-diffusion of nitrogen in UN is an important factor in the preparation and densification of UN. The self-diffusion has been measured[33] by means of ^{15}N as a tracer, and it has been found that

$$D = 2.6 \cdot 10^{-4} \exp\left(-55{,}000/RT\right) \quad (1500\text{–}1900°C)$$

Solid solutions

The solubility of various elements in UN is of special interest both for the interpretation of the physico-chemical measurements and in connection with its technological applications.

(a) Oxygen. Impurities of oxygen appear as UO (isostructural with UN) which is in solid solution with UN. The absence of a UO_2-phase is indicated by the absence of the antiferromagnetic–paramagnetic transition of UO_2 near 30°K. The maximum solubility of UO in UN has been reported by Anselin[34] to be 5 mole % between 1200–1600°C.

(b) Carbon. UN and UC are completely miscible in all ratios (see p. 173) and solid solutions of UC in UN are interesting as they bring

References p. 203

together promising properties of both UC and UN. For instance, the solid solution $U(C, N)$ has a good stability in the presence of moisture in contrast with that of pure UC. In the solid solution, the nitrogen stabilizes the UC towards graphite and thus prevents the formation of UC_2[35].

(c) *Plutonium nitride.* Recently, it was found[34] that there is complete solid solution between UN and plutonium nitride, PuN, and between UC and PuN, so that it is virtually certain that the quarternary system UC–UN–PuN–PuC forms a continuous complete solid solution $(U, Pu)(C, N)$.

As has been seen before, a suitable industrial preparation of UN is by the reaction of UO_2 with nitrogen in the presence of carbon. It is obvious that, owing to the solubility of both UO and UC in UN, a solution of the composition $UN_{1-x-y}C_xO_y$ will be obtained in this way, the composition being dependent on the pressures of nitrogen and CO employed. Imoto and Stöcker[9] have shown by thermodynamic calculation that the solid solution is non-regular. As was to be expected a high nitrogen pressure and a low CO pressure are necessary for the preparation of pure UN.

Chemical properties

The potential use of UN as a nuclear material warrents some discussion of its chemical reactivity, especially towards a number of different gases and metals. A reaction of prime importance is its behaviour upon oxidation. The reactivity of powdered uranium mononitride in respect of oxidation is primarily influenced by its surface area and particle size. UN is easily oxidized, although it is not pyrophoric except when finely divided. Fully dense, stoichiometric UN exhibits optimum oxidation resistance.

The kinetics and mechanism of the oxidation of uranium nitride has been studied by Dell *et al.*[36,37]. The oxidation mechanism follows a linear law between 230° and 300 °C with an activation energy of 30–32 kcal/mole. The oxidation rate for powdered UN is essentially independent of the oxygen pressure at these temperatures. At higher temperatures, however, a pressure-dependent oxidation rate has been observed[38].

Remarkably, the final product of the oxidation at temperatures around 250 °C is a UO_3-type phase with the composition UO_3N_x ($x = 0.2$–0.4). This phase is poorly crystalline and releases nitrogen only with difficulty, and the material then crystallizes to α-UO_3. This is followed by decomposition to U_3O_8 at 600 °C[36]. The oxidation of UN to U_3O_8 is a higly exothermic reaction (214 kcal/mole UN). The ignition characteristics of the material are therefore extremely important. The ignition temperature of UN in oxygen is a function of its specific surface area and is about 300 °C for powders sintered at 1450 °C in argon, whereas UN prepared below 1200 °C is liable to ignite on exposure to air at room temperature.

UN is stable in moist air at room temperature (in contrast to the marked reactivity of PuN). UN is even stable for days in boiling water at 100 °C. Hydrolysis of UN powder in water vapour at low pressures only occurs at an appreciable rate above 300 °C, and then according to the equation:

$$UN + 2\,H_2O \rightarrow UO_2 + NH_3 + \tfrac{1}{2}\,H_2$$

Thus, nitrogen escapes from the nitride as ammonia; nitrogen gas is not formed as has been shown mass spectrometrically by Dell et al.[39], who also have studied the kinetics of the hydrolysis of UN in water vapour.

Fully dense stoichiometric UN has, in contrast with UC, a good resistance to water corrosion at temperatures up to 300 °C[40,41]. This breakdown is because a protective oxide layer which is formed on the nitride cracks above 300 °C; after this a rapid attack follows. However, slight deviations from strict stoichiometry lead to a rapid corrosion. In hypo-stoichiometric UN the free uranium is responsible for a high corrosion rate, whereas in hyper-stoichiometric UN the corrosion is a result of a preferential attack due to the higher nitrogen potential of the U_2N_3-phase.

Uranium mononitride can be dissolved in nitric acid. The rate of dissolution is dependent on the acid concentration: in 5 N HNO_3 rapid dissolution takes place. UN is not attacked by hot HCl, H_2SO_4 and by NaOH-solutions. However, it reacts rapidly with molten alkali.

The compatibility of UN with metals at high temperatures is of considerable technological interest. In general, two types of reactions are possible: (a) nitriding $(UN + xM \rightarrow M_xN + U)$ and (b) alloying $(3\ UN + xM \rightarrow UM_x + U_2N_3)$.

The nitridation of stainless steel and other metals is a question of nitrogen potential. Pure UN is compatible with stainless steel up to at least 1200 °C. However, when impurities of U_2N_3 are present, the reaction: $U_2N_3 + 2\ Cr \rightarrow 2\ UN + Cr_2N$ may occur. Indeed, the precipitation of Cr_2N is then found at the grain boundaries.

The behaviour of zirconium and zircaloy-2 is almost identical in forming ZrN or a $(Zr,U)N$ solid solution[42]. A slow reaction occurs at 600 °C and the rate is significant at 800 °C. Aluminium reacts rapidly with UN at temperatures of 500 °C and above to form the intermetallic compound UAl_3. With nickel also a rapid reaction has been observed[43].

Sodium did not give evidence of any reaction with UN at 825 °C after a test lasting for 1000 hours. Other metals, such as chromium, molybdenum, niobium, vanadium and titanium all react with it at 1000 °C to a greater or lesser extent depending on the time. Only tungsten is reported to be compatible with UN up to 2800 °C.

Uranium nitrogen fluoride, UNF

A uranium compound containing both nitrogen and fluorine has been obtained by the reaction of UF_4 with silicon in nitrogen atmosphere at 900 °C. The material has the composition UNF[43a], which is formed by the reaction:

$$4\ UF_4 + 3\ Si + 2\ N_2 \xrightarrow{900°C} 4\ UNF + 3\ SiF_4$$

The compound UNF is obtained as a black powder. It has a tetragonal symmetry with $a = 5.612$ Å and $c = 5.712$ Å, and four molecules in the unit cell. The X-ray density is 10.01.

When UNF is heated above 1100 °C in nitrogen or argon atmosphere and in the presence of silicon, it is converted into UN.

Uranium phosphides

There are three phosphides of uranium, UP, U_3P_4 and UP_2[44]; of these the monophosphide, like UN, is of interest as a possible nuclear fuel. The compound UP is face-centered cubic (NaCl-type) with a lattice parameter of 5.589 Å; its theoretical density is 10.27. U_3P_4 is body-centered cubic (Th_3P_4-type) with a variable lattice parameter (8.21–8.25 Å), due to a considerable range of homogeneity. The cell constant of 8.207 ± 0.005 Å is reported to be that of U_3P_4 in equilibrium with UP[45].

UP_2 has a tetragonal structure with lattice parameters $a = 3.810 \pm 0.005$ Å and $c = 7.764 \pm 0.005$ Å. The cell contains two UP_2 units; the space group is P4/nmm[46].

Preparation of the uranium phosphides

The phosphides of uranium are usually prepared by the reaction of uranium with phosphine, PH_3, carried in a stream of argon. When this reaction is employed at about 400 °C with finely divided uranium, made by the decomposition of UH_3, it proceeds smoothly and easily to completion[45]. The product is UP_2; this has an appreciable phosphorus pressure above 500 °C, decomposing to the body-centered cubic U_3P_4. Any excess of phosphorus is removed by a prolonged heating of the U_3P_4 in argon at 800 °C. The latter compound itself dissociates into UP above 1100 °C, and this compound may be obtained quantitatively in vacuum at 1300 °C. The product UP is a grey powder having no range of stoichiometry at room temperature.

UP can also be prepared by direct reaction between the elements. The reaction is higly exothermic and may be carried out in an arc furnace or in a pressure vessel[47]. A disadvantage of the latter method is contamination by impurities, mainly by large amounts of oxygen in the form of UO_2.

UP_2 can be prepared by heating calculated amounts of uranium and red phosphorus at 400 °C in vacuum; the sample is then homogenized during one week at 800 °C[46].

Thermodynamic properties

The heat of formation of UP has been recently measured by fluorine-reaction calorimetry and the value $-\Delta H_{298} = 75.5 \pm 1.4$ kcal/mole was found for the reaction[27]:

$$U(\alpha) + P(\alpha\text{-white}) \rightarrow UP(s)$$

Low-temperature heat-capacity measurements have been carried out by Counsell *et al.*[48]. From their measurements it follows for UP: $S_{298}^0 = 18.71$ cal/deg.mole. A sample of U_3P_4 with the composition $P/U = 1.35$ yielded $S_{298}^0 = 20.61$ cal/deg.mole. For the heat capacity of UP between 800° and 1300 °C the value 13.8 cal/deg.mole may be used[49].

UP loses phosphorus preferentially when heated in vacuum above 1400 °C. This causes a lattice contraction with departure from stoichiometry. At still higher temperatures UP vaporizes, primarily by decomposition into gaseous uranium and phosphorus[50], leaving behind the two-phase system U (l, sat. with P) and $UP_{1-x}(c)$, in which x is between 0.02 and 0.05[51]. The partial pressures of U(g), P(g) and P_2(g) over the two-phase system have been measured[51]. The existence of a gaseous species UP at temperatures above 2270 °K has also been reported[52].

For the melting point of UP the value 2610 °C has been reported[53]. But Benz and Ward[54], who examined the U–UP system, have observed a temperature of 2850 °C for the highest melting point.

The liquidus curve in the UN–UP system has been determined by Baskin[55]. Only a slight solubility of UN in UP (about 1 mole %) has been observed, whereas complete solubility of US in UP is reported[56]. Pressure measurements in the UP_2–U_3P_4 region of the system have been made by Heimbrecht *et al.*[57].

Magnetic properties and chemical bonding

The magnetic properties of the uranium phosphides have been measured recently[58]. Both UP and UP_2 are antiferromagnetic with maximum susceptibility at the Néel point of 123 °K and 203 °K respectively. But U_3P_4 is ferromagnetic with a Curie point of 138 °K. The magnetic order in UP is the same as in UN (Néel point 52°K): the unpaired spins on the uranium atoms are ordered ferromagnetically in sheets parallel to (001) planes, the adjacent sheets being coupled antiferromagnetically[59]. The heat capacity measurements by Counsell et al.[48] exhibit a maximum C_p-value at 121 °K for UP in agreement with the magnetic data.

The magnetic properties of UX_2 (X = P, As, Sb or Bi) indicate ordering of the magnetic moments of the uranium atoms. The antiferromagnetic structure of tetragonal UP_2 (Néel temperature 203 °K) has been determined in a neutron diffraction study[46]. The magnetic unit cell of UP_2 appeared to be twice the size of the chemical one along the c-axis. Similar results have been found for the isomorphous compounds UAs_2[60] and USb_2, whereas for UBi_2 the magnetic unit cell is identical with the chemical one[61].

Low-temperature neutron diffraction measurements on UP show that uranium is in the $+3$ state with a $5f^3$ configuration[62]. The observed U–P distance in UP_2 is approximately the sum of the ionic radii of U^{4+} (0.97 Å) and P^{3-} (1.86 Å). The existence of the paramagnetic moment (2.30 B.M.) of the uranium atoms is consistent with the assumption that the oxidation state is represented by U^{4+} ions in this compound.

Chemical properties

UP is reported to be compatible with molybdenum, tantalum and tungsten up to at least 2000 °C[63]. Reaction between uranium phosphide and UO_2 was not observed up to 2500 °C.

The oxidation behaviour of UP is interesting and entirely different from that of UN. After heating UP in oxygen at 1000 °C the residue consisted of $(UO_2)_2P_2O_7$. This is explained by the formation of the two phases UO_{2+x} (or U_3O_8) and P_2O_5 in the early stage of the oxidation, followed by a further reaction to give uranyl phosphate[63].

UP only slowly dissolves in dilute mineral acids; this is accompanied by the evolution of PH_3. Sintered UP is inert to boiling water and compatible with tungsten up to 2600 °C.

Uranium arsenides and antimonides

Uranium forms with arsenic and antimony compounds similar to those formed with phosphorus[64]. Thus UAs, U_3As_4 and UAs_2 and USb, U_3Sb_4 and USb_2 have been prepared. The compounds UX_2 and U_3X_4 (X = As or Sb) are made by heating stoichiometric amounts of the powdered elements in evacuated and sealed tubes at 800 °C and 950 °C respectively.

The compounds UX are prepared by decomposing the higher pnictides in vacuum at 1300–1400 °C. A preparation of UAs has been described[65]; in this finely divided uranium (obtained via UH_3) reacts with AsH_3 gas at about 300 °C. The product has an overall As/U ratio of about 1.2, whereas X-ray analysis revealed the presence of $UAs_2 + UH_3$. The mixture, when heated in vacuum at temperatures between 1200 °C and 1400 °C, yielded uranium monoarsenide with a strong, sharp X-ray pattern with the lattice parameter $a = 5.779$ Å.

The phase diagram of the uranium–arsenic system has been examined by Benz and Tinkle[66]. From this, it appears that the highest melting point observed with the UAs-phase is 2705 °C. The solubility of uranium in the UAs-phase is small and does not exceed 0.15 wt. %.

UAs is completely soluble in UP and US[67]. Although UAs powder is not pyrophoric, it gradually oxidizes in air at room temperature. It ignites in air at about 290 °C exhibiting a sharp exothermic peak in the DTA-diagram[68].

The system uranium–bismuth

The uranium–bismuth system is of particular interest in nuclear technology. Bismuth, a low-pressure liquid with a long liquid range (m.p. = 271 °C, b.p. = 1420 °C), has a low absorption cross section for thermal neutron (0.032 b). Uranium has a good solubility in liquid bismuth according to Balzhiser[69]:

$$\log X_U = \frac{-2690}{T} + 1.215$$

Thus, a solution of uranium in liquid bismuth may act as a liquid nuclear fuel, for instance, in the thermal cycle. The activity of uranium when in dilute solution in bismuth has been measured by several authors[70,71,72]: the activity coefficient may be represented by $\log \gamma = 2.670 - 5625/T$ (773–1115 °K).

In the uranium–bismuth system the compounds UBi_2, U_3Bi_4 and UBi have been found. The system is difficult to study, since the alloys are pyrophoric and tend to form oxide layers that are not easily penetrated by X-rays. The phase diagram of the system has been determined by Teitel[73] and by Cotterill and Axon[74].

The free energy of formation of UBi, according to the equation:

$$U_{(\gamma,s)} + Bi_{(l)} \rightarrow UBi_{(s)}$$

follows from the measurements by Rice et al.[70]:

$$\Delta G_T^0 = -12,320 + 4.40\,T \quad (1000\text{–}1100\,°K).$$

REFERENCES

1 E. F. WESTRUM, JR. AND W. G. LYON, *Thermodynamics of Nuclear Materials, Proceedings of a Symposium, Vienna (1967)*, IAEA, Vienna (1968), p. 239.
2 C. RAMMELSBERG, *Pogg. Ann.*, 55 (1842) 323.
3 R. E. RUNDLE, N. C. BAENZIGER, A. S. WILSON AND R. A. McDONALD, *J. Am. Chem. Soc.*, 70 (1948) 99.
4 D. A. VAUGHAN, *J. Metals*, 206 (1956) 78.
5 C. E. PRICE AND I. H. WARREN, *Inorg. Chem.*, 4 (1965) 115.
6 W. TRZEBIATOWSKI, R. TROĆ AND J. LECIEJEWICZ, *Bull. Acad. Polon., Sci. Ser. Chim.*, 10 (1962) 395.
7 R. BENZ AND M. G. BOWMAN, *J. Am. Chem. Soc.*, 88 (1966) 264.
8 D. L. KELLER, *Euraec-Report* 556 (1963).
9 S. IMOTO AND H. J. STÖCKER, *Thermodynamics, Proceedings of a Symposium, Vienna (1965)*, IAEA, Vienna, Vol. II (1966), p. 533.
10 E. SPEIDEL AND D. L. KELLER, *BMI-Report* 1633 (1963).
11 M. ALLBUTT, A. R. JUNKISON AND R. G. GARNEY, *AERE-R-Report* 4903 (1965).
12 M. W. MALLET AND A. F. GERDS, *J. Electrochem. Soc.*, 102 (1955) 292.
13 *U. S. Patent* 3, 180, 702 (April 1965).
14 R. DIDCHENKO AND F. GORTSEMA, *Inorg. Chem.*, 2 (1963) 1079.
15 J. BUGL AND A. A. BAUER, *Compounds of Interest in Nuclear Reactor Technology, Proceedings of a Symposium (1964)*, AIME Nucl. Met. X (1964), p. 215, and in: *J. Am. Ceram. Soc.*, 47 (1964) 425.

16 W. M. OLSON AND R. N. R. MULFORD, *J. Phys. Chem.*, 167 (1963) 952.

17 T. SANO, M. KATSURA AND H. KAI, *Thermodynamics of Nuclear Materials, Proceedings of a Symposium, Vienna (1967)*, IAEA, Vienna (1968), p. 301; see also: M. KATSURA AND T. SANO, *J. Nucl. Sci. Techn.*, 4 [6] (1967) 283.

18 P. A. VOZELLA AND M. A. DECRESCENTE, *PWAC-Report* 479 (1965).

19 H. INOUYE AND J. M. LEITNAKER, *J. Am. Ceram. Soc.*, 51 (1968) 6.

20 E. H. P. CORDFUNKE AND K. A. NATER, to be published.

21 P. E. LAPAT AND R. B. HOLDEN, *Compounds of Interest in Nuclear Reactor Technology, Proceedings of a Symposium (1964)*, AIME Nucl. Met. X (1964), p. 225.

22 F. MÜLLER AND H. RAGOSS, *Thermodynamics of Nuclear Materials, Proceedings of a Symposium, Vienna (1967)*, IAEA, Vienna (1968), p. 257.

23 M. H. RAND, private communication.

24 A. NAOUMIDIS, *Thesis*, Jül-472-RW (1967). See also H. J. STÖCKER AND A. NAOUMIDIS, *Ber. Deutsch. Keram. Gesellsch.*, 43 (1966) 724.

25 N. C. BAENZIGER, *CC-Report* 1984 (1944), quoted in: J. J. KATZ AND E. RABINOWITCH, *The Chemistry of Uranium*, Part I, McGraw-Hill Book Co., Inc., New York (1951), p. 239.

26 P. GROSS, C. HAYMAN AND H. CLAYTON, *Thermodynamics of Nuclear Materials, Proceedings of a Symposium, Vienna (1962)*, IAEA, Vienna (1963), p. 653.

27 P. A. G. O'HARE, J. L. SETTLE, H. M. FEDER AND W. N. HUBBARD, *Thermodynamics of Nuclear Materials, Proceedings of a Symposium, Vienna (1967)*, IAEA, Vienna (1968), p. 265.

28 E. F. WESTRUM, JR. AND CAROLYN M. BARBER, *J. Chem. Phys.*, 45 (1966) 635.

29 J. F. COUNSELL, R. M. DELL AND J. F. MARTIN, *Trans. Faraday Soc.*, 62 (1966) 1736.

30 M. ALBUTT, A. R. JUNKISON AND R. M. DELL, *Compounds of Interest in Nuclear Reactor Technology, Proceedings of a Symposium (1964)*, AIME Nucl. Met. X (1964), p. 65.

31 W. TRZEBIATOWSKI AND R. TROĆ, *Bull. Acad. Polon., Sci. Ser. Chim.*, 12 (1964) 681.

32 D. L. KELLER, *Euraec-Report* 77 (1961).

33 T. J. STURIALE AND M. A. DECRESCENTE, *PWAC-Report* 477 (1965).

34 F. ANSELIN, *CEA-Report* 2988 (1966).

35 M. RAND, *AERE-M-Report* 1360 (1964).

36 R. M. DELL, V. J. WHEELER AND E. J. MCIVER, *Trans. Faraday Soc.*, 62 (1966) 3591.

37 R. M. DELL AND V. J. WHEELER, *J. Nucl. Mat.*, 21 (1967) 330.

38 J. BESSON, C. MOREAU AND J. PHILIPPOT, *Bull. Soc. Chim. France*, (1964) 1069.

39 R. M. DELL, V. J. WHEELER AND N. J. BRIDGER, *Trans. Faraday Soc.*, 63 (1967) 1286.

40 J. BUGL AND A. A. BAUER, *Compounds of Interest in Nuclear Reactor Technology, Proceedings of a Symposium (1964)*, AIME Nucl. Met. X (1964), p. 463.

41 J. E. ANTILL AND B. C. MYATT, *Corrosion Sci.*, 6 (1966) 17.

42 D. E. PRICE, D. P. MOAK, W. CHUBB AND D. L. KELLER, *BMI-Report* 1760 (1966).

43 M. ALLBUTT AND A. R. JUNKISON, *AERE-R-Report* 5466 (1967).

43a K. YOSHIHARA, M. KANNO AND T. MUKAIBO, *J. Inorg. Nucl. Chem.*, 31 (1969) 985.

44 M. HEIMBRECHT AND M. ZUMBUSCH, *Z. Anorg. Allgem. Chem.*, 245 (1941) 391.

45 Y. BASKIN AND P. D. SHALEK, *J. Inorg. Nucl. Chem.*, 26 (1964) 1679.

46 R. TROĆ, J. LECIEJEWICZ AND R. CISZEWSKI, *Phys. Stat. Sol.*, 15 (1966) 515.

47 J. B. MOSER AND O. L. KRUGER, *Ceramic Bull.*, 12 (10) (1963) 61.

48 J. F. COUNSELL, R. M. DELL, A. R. JUNKISON AND J. F. MARTIN, *Trans. Faraday Soc.*, 63 (1967) 72.

49 *ANL-Report* 7225 (1966).

50 K. A. GINGERICH AND P. K. LEE, *J. Chem. Phys.*, 40 (1964) 3520.

51 *ANL-Reports* 7375 (1967) and 7450 (1968).

52 K. A. GINGERICH, *Naturwissensch.*, 53 (20) (1966) 525.

53 Y. BASKIN, *Nucl. Sci. Eng.*, 24 (4) (1966) 332.

54 R. BENZ AND C. H. WARD, *J. Inorg. Nucl. Chem.*, 30 (1968) 1187.

55 Y. BASKIN, *J. Am. Ceram. Soc.*, 50 (1967) 74.

56 P. D. SHALEK AND Y. BASKIN, *Bull. Am. Ceram. Soc.*, 43 (4) (1964) 327.

57 M. HEIMBRECHT, M. ZUMBUSCH AND W. BILTZ, *Z. Anorg. Allgem. Chem.*, 245 (1940) 391.

58 W. TRZEBIATOWSKI AND R. TROĆ, *Bull. Acad. Polon., Ser. Sci. Chim.*, 11 (1963) 661.

59 N. A. CURRY, *Proc. Phys. Soc.*, 86 (1965) 1193.

60 A. OLÈS, *J. Phys.*, 26 (1965) 561.

61 J. LECIEJEWICZ, R. TROĆ, A. MURASIK AND A. ZYGMUND, *Phys. Stat. Sol.*, 22 (1967) 517.

62 S. S. SIDHU, W. VOGELSANG AND K. D. ANDERSON, *J. Phys. Chem. Solids*, 27 (1966) 1197.

63 Y. BASKIN AND P. D. SHALEK, *Compounds of Interest in Nuclear Reactor Technology, Proceedings of a Symposium (1964)*, AIME Nucl. Met. X (1964), p. 457.

64 W. TRZEBIATOWSKI, A. SEPICHOWSKA AND A. ZYGMUNT. *Bull. Acad. Polon., Ser. Sci. Chim.*, 12 (1964) 687.

65 Y. BASKIN, *J. Inorg. Nucl. Chem.*, 29 (1967) 2480.

66 R. BENZ AND M. C. TINKLE, *J. Electrochem. Soc.*, 115 (1968) 322.

67 *ANL-Report* 7249 (1966).

68 Y. BASKIN, *J. Am. Ceram. Soc.*, 48 (1965) 153.

69 R. E. BALZHISER AND D. V. RAGONE, *Trans. Met. Soc. AIME*, 224 (1962) 485.

70 P. A. RICE, R. E. BALZHISER AND D. V. RAGONE, *Thermodynamics of Nuclear Materials, Proceedings of a symposium, Vienna (1962)*, IAEA, Vienna (1963), p. 331.

71 R. H. WISWALL AND J. J. EGAN, *ibid.*, p. 345.

72 L. C. TIEN, K. J. GUION AND R. D. PEHLKE, *Thermodynamics, Proceedings of a Symposium, Vienna (1965)*, IAEA, Vienna, Vol. I (1966), p. 501.

73 R. J. TEITEL, *J. Metals*, 9 (1957) 131.

74 P. COTTERILL AND H. J. AXON, *J. Inst. Metals*, 87 (1959) 159.

CHAPTER 12

Chalcogenides of Uranium

Uranium sulphides

Several investigations of the phases in the uranium–sulphur system had been made[1,2,3,4], but the results appeared to be rather contradictory until Picon and Flahaut[5], in a comprehensive study, showed the existence of the phases US, U_2S_3, U_3S_5, US_2 (α, β and γ) and US_3. These authors also studied the principal properties of these phases so that most of our present knowledge of the uranium–sulphur system stems from their work. Only the properties of the monosulphide US have received more detailed attention because of its interest as a potential nuclear fuel for power reactors.

Crystallographic properties. Phase diagram

The crystallographic properties of the different phases in the uranium–sulphur system are listed in Table 23. It should be noted that the lattice parameter of US varies between 5.470 Å and 5.493 Å. The lower values are reported[6] to be associated with diffuse X-ray patterns and generally with high oxygen contents. The value 5.468 Å is the lattice parameter of US at room temperature after it has been saturated with oxygen at high temperatures (about 1 mole %). Moreover, lattice contraction takes place with a departure from stoichiometry (sulphur deficiency).

The phase relationships are not definitely settled. Nevertheless, the tentative phase diagram shown in Fig. 37 may be used. Near the region S/U = 1.0, a narrow composition range may exist; the maximum limits are 0.96 and 1.01[7]. The melting point of US is $2462 \pm 30\,°C$. The stoichiometric phases U_2S_3 and U_3S_5 decompose at 1950 °C and 1800 °C respectively.

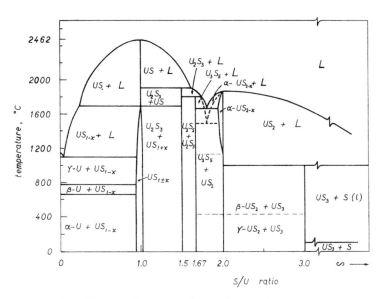

Fig. 37. Phase diagram of the uranium–sulphur system.

According to Picon and Flahaut[5] the disulphide has three crystallographic forms of which γ-US$_2$ is the low-temperature form. The transition of γ-US$_2$ to β-US$_2$ takes place at about 425 °C; the high-temperature form, α-US$_2$, is produced at $1350° \pm 20$ °C:

$$\gamma\text{-US}_2 \xrightarrow{425\ °C} \beta\text{-US}_2 \xrightarrow{1350\ °C} \alpha\text{-US}_2$$

However, Grønvold et al.[8] state that α-US$_2$ has a rather broad homogeneity range between US$_{1.8}$ and US$_{1.95}$; it should therefore not be considered as a polymorph of the disulphide. At these two compositions the lattice constants of the tetragonal phase are $a = 10.301$ Å, $c = 6.363$ Å and $a = 10.259$ Å, $c = 6.342$ Å, respectively. These authors could not confirm the formation of γ-US$_2$: even after heat treating US$_2$ at 400–410 °C for one month, no γ-US$_2$ could be observed. The β-US$_2$ phase has a composition close to stoichiometry.

The phase U$_3$S$_5$ was obtained free from other sulphide phases only at the composition US$_{1.60}$.

References p. 216

TABLE 23

CRYSTALLOGRAPHIC PROPERTIES OF THE URANIUM–SULPHUR PHASES

Phase	Symmetry	Lattice parameters ($\mathring{A}$)		Density	Ref.
US	cubic (NaCl-type)	$a =$	5.486		5
			5.473		2
			5.4903	10.87	7
U_2S_3	orthorhombic	$a =$	10.36	8.96	5
		$b =$	10.60		
		$c =$	3.863		
U_3S_5	orthorhombic	$a =$	7.42		
		$b =$	8.08	8.26	5
		$c =$	11.72		
α-US$_2$ ($= US_{1.8-1.95}$)	tetragonal	$a =$	10.301–10.259	7.5	8
		$c =$	6.342– 6.363		
β-US$_2$	orthorhombic	$a =$	4.13		
		$b =$	7.12	8.03	5
		$c =$	8.48		
γ-US$_2$	hexagonal	$a =$	7.252		
		$c =$	4.067	8.126	5
US$_3$	monoclinic	$a =$	5.40		
		$b =$	3.90	5.86	5
		$c =$	18.26		
		$\alpha =$	80°30′		

Thermodynamic properties

The heat of formation of US has been determined recently[9]; a value of $-\Delta H_{298} = 73.2 \pm 3.6$ kcal/mole has been found. The entropy of US has the value $S_{298}^0 = 18.64$ cal/deg.mole as determined from low-temperature heat capacity measurements[10,11]. The thermodynamic functions of US, β-US$_2$ and US$_3$ below 350°K have been given by Westrum *et al.*[11,12].

The high-temperature heat capacity of the monosulphide has been measured recently[13]. From the heat contents:

$$H_T - H_{298} = 12.633\,T + 7.7865 \cdot 10^{-4}\,T^2 + 9.0413 \cdot 10^4\,T^{-1} - 4139$$

$$(298-1400°K)$$

it follows for the heat capacity of US:

$$C_p = 12.633 + 15.573 \cdot 10^{-4} T - 9.0413 \cdot 10^4 T^{-2}$$

The dissociation pressure in the change of US_3 to US_2 and sulphur has been measured by Strotzer et al.[1]; the heat of formation of US_3 was calculated from the measurements to be $-\Delta H_{298} = 125$ kcal/mole.

Uranium monosulphide vaporizes congruently to $US_{(g)}$, $U(g)$ and $S_2(g)$. Using an effusion cell, Cater[7] studied the evaporation. The vapour pressure and mass spectrometric results were used to separate the contribution of the gaseous uranium and monosulphide species, and to give the free energies of formation[14,15,16]. From the results, it follows that the ratio of the partial pressures of $US(g)$ and $U(g)$ at $2000\,°K$ is 0.27. The results, which are in good agreement with recent measurements by Nater[17], are

for US(g): $\qquad \log p(\text{atm}) = \dfrac{-31,030}{T} + 7.606$

for U(g): $\qquad \log p(\text{atm}) = \dfrac{-29,600}{T} + 7.323$

It has been suggested recently[17a] that the compound $US_{1.00}$ does not vaporize congruently, but changes in composition within a homogeneity range towards US_{1-x} (with $x = f(T)$), until the phase limit is reached; this would occur at $2350\,°K$.

To summarize results from the different investigations, the values of the relevant thermodynamic properties are listed in Table 24.

TABLE 24

THERMODYNAMIC PROPERTIES OF URANIUM SULPHIDE PHASES

	US	U_2S_3	$US_2(\beta)$	US_3
S_{298}^0 (cal/deg.mole)	18.64	46.6	26.39	33.10
$-\Delta H_{298}$ (kcal/mole)	73.2	224	120	125
C_p (cal/deg.mole)	12.08	—	17.84	22.85

Preparation of the uranium sulphides

Uranium monosulphide is prepared from bulk uranium by hydriding it to UH_3, followed by reaction of the finely divided uranium thus obtained with an equivalent quantity of purified hydrogen sulphide at about 500 °C[3, 6, 18]. After completion of the sulphiding the product is annealed at 2000 °C either in argon or a good vacuum (10^{-6} mm) to complete homogenization. The annealed monosulphide is bright, silvery and of metallic appearance.

US may also be prepared in a massive form by argon arc melting. Up to the present, reductions of UO_2 to US have not been very successful.

The sinterability of US powders is strongly affected by the presence of secondary phases. Small amounts of the oxysulphide UOS aided the densification through the mechanism of liquid phase sintering. This resulted in samples of a density as high as 95 % theoretical. High-purity US, however, could only be sintered to about 85 % theoretical density[18].

Uranium sulphides can be synthesized by the reaction of metallic uranium with elemental sulphur in the appropriate ratio in small alumina crucibles inside evacuated and sealed quartz tubes. The preparation has to be done very carefully[8]. The uranium sulphides are all rather unstable in air and all handling of the materials has to be carried out in an inert atmosphere dry box.

Special methods for the preparation of the different forms of US_2 are now described[5, 6].

γ-US_2 is formed as a surface layer on uranium metal by the reaction of H_2S at about 400 °C and may be prepared in mass by the reaction of U_3S_5 (see below) with sulphur at about 400 °C.

β-US_2 is prepared either by the reaction of UCl_4 with H_2S at 600 °C or by that of uranium with H_2S at 800–900 °C. Single crystals of β-US_2 have been prepared by allowing gaseous uranium iodide and sulphur to react 950 °C[19]. Small single crystals of β-US_2 are then deposited in the cooler parts of the apparatus (700 °C).

α-US_2 is obtained when UO_2 or U_3O_8 react with H_2S at a high temperature. The reaction begins at about 1000 °C and proceeds rapidly above 1200 °C.

The preparation of the other sulphides usually starts from the disulphide. Thus, US_3 may be made[5] by heating US_2 and an excess of sulphur in a sealed tube at 400–600 °C until the reaction is completed. The excess of sulphur is removed by washing with CS_2; the product is dried in vacuum.

The sulphide U_2S_3 results from the reduction of US_2 with aluminium powder, initially at 1200 °C, finally at 1325 °C and a pressure of 10^{-3} mm. The Al_2S_3 and the excess of aluminium are removed by treating the product with boiling acetic acid.

The sulphide U_3S_5 can be prepared in two ways: (a) by reducing α-US_2 with hydrogen at a temperature of about 1500 °C, and (b) by the dissociation of US_2 in vacuum (10^{-3} mm) at 1500 °C. Picon and Flahaut[5] have shown that the final product in both reactions is U_3S_5 which is quite free of the presence of U_2S_3.

Chemical properties of the uranium sulphides

Uranium monosulphide is not attacked by boiling water. In dilute, oxidizing acids it dissolves with the liberation of H_2S. However, treatment with HCl of any strength is without effect. US has a good resistance to oxidation up to 300 °C; then a slow reaction producing uranium oxysulphide, UOS, takes place. In powder form it ignites at 360–375 °C. The DTA-diagrams show three exothermic peaks in rapid succession due to a stepwise oxidation of US via UOS to UO_{2+x}[18]. Kolar[20], however, using DTA-analysis showed that the oxidation of US is, under proper reaction conditions, a simple one-step reaction. The main reaction, the oxidation of US to UO_2, finishes at 400 °C. At this temperature some U_3O_8 is already present; at 500 °C the only solid reaction product is U_3O_8.

US is compatible with virtually all possible cladding materials over a wide temperature range. Thus, molybdenum, niobium, tantalum and vanadium do not react with US up to 2000 °C, and even in the presence of sodium, reaction was not observed with these metals up to 800 °C.

The chemical properties of the other sulphides are rather similar. They remain resistant to oxygen up to 300 °C and are not attacked by mineral acids, unless these are oxidizing. The sulphides are resistant to alkaline solutions.

Physical properties

The thermal conductivity of US at 75°C is 0.0264 cal/sec.cm deg.[18,21] and thus is about 35 % higher than the value for UO_2. At 100°C the thermal conductivity is 0.027, raising to 0.0408 at 1000°C. It should be noted that 80 % dense US has a significantly lower thermal conductivity than sulphides with a density 95 % of theoretical.

The electrical resistivity of US ranged from 112 to 360 μ Ωcm and shows a small, but significant increase with temperature as could be expected from the metallic nature of this compound.

US is a soft material with a Vickers hardness of 200 kg/mm^2. Irradiation experiments have shown that high-density, liquid-phase, sintered US has a good irradiation stability and releases little fission gas[22].

Not much has been reported on the physical properties of the sulphides with a higher sulphur content.

Chemical bonding in the uranium sulphides

Much information concerning the chemical bonding in the uranium sulphides has been derived from magnetic measurements. Thus, Picon and Flahaut[5] have shown that every sulphide of uranium is paramagnetic at room temperature. The strong paramagnetism, observed for $US_3 (\mu_B = 2.5$ B.M.) indicates that this sulphide does not contain diamagnetic U^{6+} ions; thus it should be considered as a polysulphide.

β-US_2 has a magnetic moment (2.85 B.M.) close to the spin-only moment for two unpaired electrons ($= 2.83$ B.M.). From this, Picon and Flahaut[5] assumed the uranium to be in the tetravalent state, the two unpaired electrons being in the 6d-orbital. However, Grønvold *et al.*[8] found 3.03 B.M. for the magnetic moment of US_2; this casts some doubt on this assumption. US_2 shows no thermal anomaly in the low-temperature heat capacity curve[10].

US is ferromagnetic below 180°K; at this temperature disordering of the ferromagnetic state takes place with a transitional entropy increment of 1.17 cal/deg.mole[11]. In US the presence of U^{4+} ions is assumed[5]. Two d electrons serve to ionize the sulphur atom and two others are used in strong metal-bonds (evidenced by the high electrical conductivity), thus leaving uranium in the $+4$ state.

Solid solutions

The important potential nuclear fuel materials US, UC and UN all have an NaCl-type structure. Their solid solutions are interesting since they might embody the best features of the individual compounds. The solubility of the different phases in US is limited by consideration of ionic radii. Thus, UN ($a = 4.89$ Å) has a solubility in US ($a = 5.49$ Å) of only 10 mole % at $1825\,°C$[13].

UC ($a = 4.96$ Å) and US are only partially miscible[21]; the limit of solubility of UC in US is 40 mole % at $1600\,°C$. The maximum solubility of US in UC is about 4 mole % at $1600\,°C$. The US–UC system contains no intermediate compounds; the liquidus exhibits a maximum at 40 mole % and $2540\,°C$.

UP ($a = 5.59$ Å) and US are completely miscible; a maximum melting point of $2666\,°C$ was observed at 70 mole % UP[21]. The system US–UP is interesting since it is one of the few systems in which there is a complete solubility between a ferromagnetic (US) and an antiferromagnetic compound. Low-temperature heat capacities of different compositions in the solid solution range have been measured[23]; from this it is found that as P is progressively substituted for S in US, the magnetic ordering temperature decreases. Clearly, the ferromagnetic interactions predominate over the antiferromagnetic interactions in this system up to about 87 mole % UP; then antiferromagnetism was found.

Uranium oxysulphide

The oxysulphide of uranium, UOS, was first discovered by Eastman et al.[3], who prepared it by passing H_2S over UO_2 at about $1000\,°C$, that is below the temperature required for the formation of US. The oxysulphide may also be prepared by the reaction of equivalent molecular amounts of UO_2 and US_2 at 1350°C in vacuum[5]. UOS has a hexagonal crystal structure (PbCl-type), with lattice parameters $a = 3.841$ Å and $c = 6.694$ Å; its X-ray density is 9.644. The ordering of the magnetic moments in the antiferromagnetic structure (Néel temperature is 55°K) was determined by means of neutron diffraction[24].

UOS is the principal contaminant in the preparations of US. For instance, in a vacuum of 10^{-4} mm the monosulphide is oxidized to UOS and UO_2 when heated at high temperatures. However, UOS appears to be more volatile than US, and it disappears when the material is further heated in a very good vacuum[15]. The phase relationships between US, UOS and UO_2 are thus of much interest. The partial pressure of oxygen over this three-phase system has been determined as a function of the temperature in the range 1050–1350°K[25]. It was found:

$$\log p_{O_2}(\text{atm}) = \frac{-14,700}{T} - 10.5$$

A mass spectrometric study of the $US_2 - UO_2$ system[26] has shown that the vapour species at 2130°K are U, UO, UO_2, UOS, US, S, and presumably O, with UO being the predominant species.

A eutectic occurs in the system at 2240°K and at approximately 45% UO_2.

Uranium selenides

In the uranium–selenium system the phases USe, U_3Se_5, U_2Se_3, USe_2 and USe_3 have been described[27,28,29]. U_3Se_5 is the most uranium-rich selenide; it has an orthorhombic structure, probably isomorphous with U_3S_5. Uranium diselenide is reported to occur in three modifications (α, β, γ), but the α-USe_2 phase was found recently[8] to have a variable composition (like $US_{1.9}$) and should be considered therefore as a separate phase. It has a tetragonal structure with $a = 10.772$ Å and $c = 6.668$ Å for the composition $US_{1.8}$. The existence of the earlier reported phases β-US_2 and γ-US_2 could not be confirmed[8]. The only phase richer in selenium than $US_{1.9}$ found was the monoclinic USe_3-phase with the lattice constants $a = 5.65$ Å, $b = 4.06$ Å, $c = 9.55$ Å, and $\beta = 97.5°$. USe has been reported to have the NaCl-structure.

Preparation

The preparation of USe_3 may be carried out by the reaction of H_2Se on either UCl_4 or metallic uranium. The phases with a lower selenium content may be made by heating USe_3 to between 500–1500 °C in high vacuum.

Properties of USe

Uranium monoselenide is cubic (NaCl-type) with a lattice constant of 5.71 Å and an X-ray density of 11.3[30]. USe is of particular interest because of its ferromagnetism below 185–190 °K[31], and its low electrical conductivity and high thermoelectric power at room temperature, actually $+ 38$ $\mu V/°C$[30]. At high temperatures the thermoelectric properties of USe may be found to compare favourably with the best known thermoelectric materials.

The low-temperature heat capacity curve of USe shows a λ-type transition at 160.5 °K to be associated with ferromagnetic ordering, although the transition temperature is 25–30° lower than the Curie point obtained from magnetic measurements[31]. At room temperature the entropy of USe has the value $S^0_{298} = 23.07$ cal/deg.mole; the heat capacity $C_p = 13.10$ cal/deg.mole.

USe reacts readily with acids, hydrogen selenide being generated. Neither USe nor U_2Se_3 was found to react with 3 N KOH[30]. USe is stable in air, at 300 °C only a surface reaction has been observed.

USe_2 shows a small λ-type anomaly at 13.1 °K[32], presumably due to magnetic ordering at this temperature. The entropy determined from these measurements has the value $S^0_{298} = 32.0$ cal/deg.mole.

Uranium tellurides

In the uranium–tellurium system, phases analogous with the uranium selenides have been identified. Thus UTe, U_3Te_4, U_2Te_3,

$UTe_2(\alpha, \beta, \gamma)$ and UTe_3 have been reported[27,30]. Uranium mono-telluride was found to be polyphase; samples of it consist of a major cubic phase with the Th_3P_4 structure ($a = 9.393$ Å; and a density of 8.81) and a minor phase with NaCl-type structure ($a = 6.151$ Å and a density of 10.43). The Th_3P_4 structure was also observed for U_3Te_4 and U_2Te_3[30]. UTe_3 is isomorphous with the trisulphide and triselenide; it is monoclinic with $a = 6.10$ Å, $b = 4.22$ Å, $c = 10.34$ Å, and $\beta = 97°8'$[8].

The electrical properties of the tellurides of uranium have been measured at various temperatures[30]. In addition, the magnetic properties of these phases have also been measured[33,34], from these it appears that UTe is weakly ferromagnetic below 104 °K. The entropy of UTe at room temperature has the value $S^0_{298} = 26.7$ cal/deg.mole[10].

UOTe has a similar magnetic structure as UOS[35]; the ordering of the magnetic moments in the antiferromagnetic structure has been studied by neutron diffraction.

REFERENCES

1 E. F. STROTZER, O. SCHNEIDER AND W. BILTZ, Z. Anorg. Allgem. Chem., 243 (1940) 297.
2 W. H. ZACHARIASEN, Acta Cryst., 2 (1949) 288.
3 E. D. EASTMAN, L. BREWER, L. A. BROMLEY, P. W. GILLES AND N. L. LOF-GREN, J. Am. Chem. Soc., 72 (1950) 4019.
4 E. D. EASTMAN, L. BREWER, L. A. BROMLEY, P. W. GILLES AND N. L. LOF-GREN, J. Am. Chem. Soc., 73 (1951) 3896.
5 M. PICON AND J. FLAHAUT, Bull. Soc. Chim. France (1958) 772.
6 M. ALLBUTT, A. R. JUNKISON AND R. G. CARNEY, AERE-R-Report 4903 (1965).
7 E. D. CATER, ANL-Report 6140 (1960).
8 F. GRØNVOLD, H. HARALDSEN, T. THURMANN-MOE AND T. TUFTE, J. Inorg. Nucl. Chem., 30 (1968) 2117.
9 P. A. G. O'HARE, J. L. SETTLE, H. M. FEDER AND W. N. HUBBARD, Thermodynamics of Nuclear Materials, Proceedings of a Symposium, Vienna (1967), IAEA, Vienna (1968), p. 265.
10 E. F. WESTRUM, JR. AND F. GRØNVOLD, Thermodynamics of Nuclear Materials, Proceedings of a Symposium, Vienna (1962), IAEA, Vienna (1963), p. 3.
11 E. F. WESTRUM, JR., R. R. WALTERS, H. E. FLOTOV AND D. W. OSBORNE, J. Chem. Phys., 48 (1968) 155.

12 F. GRØNVOLD AND E. F. WESTRUM, JR., *J. Inorg. Nucl. Chem.*, 30 (1968) 2127.

13 A. C. MACLEOD AND S. W. J. HOPKINS, *Proc. Brit. Ceram. Soc.*, 8 (1967) 15.

14 E. D. CATER, P. W. GILLES AND R. J. THORN, *J. Chem. Phys.*, 35 (1961) 608.

15 E. D. CATER, E. G. RAUH AND R. J. THORN, *J. Chem. Phys.*, 35 (1961) 619.

16 E. D. CATER, *J. Chem. Phys.*, 44 (1966) 3106.

17 K. A. NATER, to be published.

17a A. PATTORET, J. DROWART AND S. SMOES, *Trans. Faraday Soc.*, 65 (1969) 98.

18 P. D. SHALEK, *J. Am. Ceram. Soc.*, 46 (1963) 155.

19 P. K. SMITH AND L. CATHEY, *J. Electrochem. Soc.*, 114 (1967) 973.

20 D. KOLAR, M. KOMAC, M. DROFENIK, V. MARINKOVIC AND N. VENE, *Thermodynamics of Nuclear Materials, Proceedings of a Symposium, Vienna (1967)*, IAEA, Vienna (1968), p. 279.

21 Y. BASKIN AND P. D. SHALEK, *Compounds of Interest in Nuclear Reactor Technology, Proceedings of a Symposium (1964)*, AIME Nuclear Metallurgy, Vol. X (1964), p. 457.

22 L. A. NEIMARK AND R. CARLANDER, *ibid.*, p. 753.

23 J. F. COUNSELL, R. M. DELL, A. R. JUNKISON AND J. F. MARTIN, *Thermodynamics of Nuclear Materials, Proceedings of a symposium, Vienna (1967)*, IAEA, Vienna (1968), p. 385.

24 R. BALLESTRACCI, E. F. BERTAUT AND R. PAUTHENET, *J. Phys. Chem. Solids*, 24 (1963) 487.

25 W. A. YUILL, JR. AND E. D. CATER, *J. Phys. Chem.*, 71 (1967) 1436.

26 E. D. CATER, E. G. RAUH AND R. J. THORN, *J. Chem. Phys.*, 49 (1968) 5244.

27 R. FERRO, *Z. Anorg. Allgem. Chem.*, 275 (1954) 320.

28 P. KHODADAD, *Compt. Rend.*, 250 (1960) 3998.

29 P. KHODADAD, *Bull. Soc. Chim. France* (1961) 133.

30 L. K. MATSON, J. W. MOODY AND R. C. HIMES, *J. Inorg. Nucl. Chem.*, 25 (1963) 795.

31 W. TRZEBIATOWSKI AND W. SUSKI, *Bull. Acad. Polon., Ser. Sci. Chim.*, 10 (1962) 399.

32 Y. TAKAHASHI AND E. F. WESTRUM, JR., *J. Phys. Chem.*, 69 (1965) 3618.

33 W. TRZEBIATOWSKI AND A. SEPICHOWSKA, *Bull. Acad. Polon., Ser. Sci. Chem.*, 7 (1959) 181.

34 W. TRZEBIATOWSKI, J. NIEMEC AND A. SEPICHOWSKA, *ibid.*, 9 (1961) 373.

35 A. MURAKSI AND J. NIEMEC, *ibid.*, 13 (1965) 291.

CHAPTER 13

Some Applications of Uranium Chemistry in Nuclear Technology

Introduction

Uranium chemistry is concerned with the preparation and properties of uranium compounds. Many of these have been described in previous chapters, and their relevance to the problems that have arisen in the development of nuclear technology have been indicated there. Besides the materials U, UO_2 and UC, which have found direct application as fuels in the nuclear reactor, there are many other uranium compounds involved in this development and their chemical behaviour and physical properties are of key importance for this reason.

Special attention has been paid to thermochemical properties since the most valuable application of this information is prediction of the stability of particular phases in temperature gradients and of the compatibility of materials. In addition, an accurate knowledge of phase diagrams is vital for indication of the expected limits between which different materials can be used and also for the control of their production.

In this chapter we attempt to summarize the properties of uranium compounds of special interest in nuclear technology and to make certain intercomparisons. This may indicate some of the many problems with which uranium chemistry is concerned in nuclear technology.

Metallic or ceramic nuclear fuel?

Uranium metal was for long the only nuclear fuel since it is the material with the highest uranium density, and compared with other types of uranium fuels, it has relatively good physical properties. It has been used, notwithstanding several recognized limitations and drawbacks, such as phase transformations at relatively low temperatures ($\alpha \rightarrow \beta$ at 668 °C) and dimensional instability during irradiation which may occur either by anisotropic growth or by high-temperature swelling. As described in the Chapters 3 (p. 36) and 4 (p. 54), numerous attempts have been made to minimize these effects by preparing uranium alloys which are crystallographically stable and which have controlled grain size and orientation. In these alloys swelling is less and corrosion resistance and strength is better than in the unalloyed metal. By these means considerable progress has been made towards a better understanding and control of irradiation effects. Indeed, anisotropic growth in metallic fuels is no longer a serious technological problem and, in consequence, metallic fuels are still frequently used.

Provided that swelling can be kept within certain limits, the maximum temperature at which metallic uranium may be used seems to be somewhere between 500° and 650 °C. But the failure of solid metal fuels to operate successfully at high fission rates or at temperatures beyond this range, as is required in power reactors, has necessitated a search for alternative fuels. This has led to much work since 1955 on possible ceramic alternatives, especially on UO_2.

The important question is: what properties should be looked for in the ideal ceramic fuel? Obviously, and most importantly, the material needs a high concentration of uranium per unit mass and a high melting point that can be reached without phase change. Further, it has to meet stringent requirements in respect of thermal conductivity, chemical compatibility with cladding materials, and ability to withstand high irradiation doses.

Such requirements are more or less satisfied by such materials as uranium dioxide with a cubic, fluorite-type structure, and by the semi-metallic, NaCl-type, monocarbide, nitride, phosphide and sulphide of uranium.

At present, the non-metallic fuels most commonly employed are ceramic uranium dioxide, and, for fast reactors employing plutonium, a solid solution of the dioxides of uranium and plutonium, $(U, Pu)O_2$. The evident economic importance of this fuel has initiated an incredible amount of both fundamental and technological research. Properties, such as ease of fabrication, dimensional stability, fission product retention, and thermal conductivity, both out- and in-pile, have been examined in much detail. The determination of thermodynamic properties and phase relationships at very high temperatures appeared to be necessary and have in turn stimulated the development of quite new techniques of investigation.

A case in point is the accurate measurement of free energies of the UO_{2-x} phase by means of EMF-measurements at temperatures around $2000\,°C$[1]. Vaporization processes are a subject of equal importance and have been examined for UO_{2-x} by means of mass spectrometry. Incidentally, the volatile oxides UO and UO_2 have been detected in this part of the phase diagram and their vapour pressures have been measured.

However, the oxide-fuels, although well-established now, possess two inherent limitations, namely a comparatively low uranium density and poor thermal conductivity. The latter gives rise to thermal stresses because of the large temperature gradient from the centre to the outside of the fuel element during operation. This uneven distribution of temperature leads to microstructural changes (grain growth, and the formation of columnar grains) in the fuel, which in turn alter the conductivity characteristics, probably for the worse. Shortly after the fuel element has reached its maximum output of power, it takes on a characteristic structure, which it will in all essentials maintain during its useful life in the reactor. The heat transfer properties it then has, will, however, be different from those calculated on the basis of the original density.

The disadvantage of ceramic UO_2, especially for use in fast reactors where the heat flux is high, have been an incentive to the investigation of the NaCl-type, semi-metallic uranium compounds as alternative fuels.

Semi-metallic uranium fuels

The group of semi-metallic, NaCl-type uranium compounds, for instance, the monocarbide, nitride, phosphide and sulphide, have properties at high-temperatures (Table 25) which make them suitable for use as reactor fuels. They have uranium densities, much higher than that of UO_2; they are good electrical conductors and have a good thermal conductivity. The best-known compound of this group is uranium monocarbide, UC, which has long been recognized for its potential advantage over the metal and the oxide, because of (a) its superior behaviour under irradiation, to that of metallic fuel, and (b) its better thermal conductivity (Fig. 38),

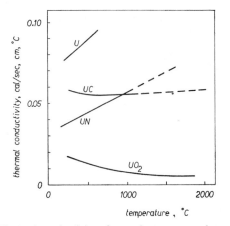

Fig. 38. Thermal conductivity of some important nuclear materials.

higher fissile density and lower fission-gas release than the oxide. Therefore, interest has largely centred upon the carbides UC and PuC, and their solid solutions. However, these compounds possess one serious drawback, namely incompatibility with cladding materials; for instance, their compatibility with stainless steel is poor. Especially is this the case with hypostoichiometric UC which cannot be used because of its objectionable behaviour with the cladding metal. A better choice may be hyperstoichiometric uranium monocarbide, UC_{1+x}; but this causes carburization of the cladding, however the addition of alloying elements, such as molybdenum, chromium or vanadium (1–3 %), has been successfully used to remove the excess of carbon in the carbide, produced at a high burn-up.

TABLE 25

SOME PHYSICAL PROPERTIES OF CERAMIC URANIUM COMPOUNDS

Compound	Melting point (°C)	Density	Uranium density	Structure	Thermal conductivity (cal/cm · sec · deg)	
					25°C	1000°C
UO_2	2800°	10.95	9.6	cubic, CaF_2-type, $a = 5.470$ Å	0.024	0.007
UC	2525°	13.63	12.97	cubic, NaCl-type, $a = 4.960$ Å	0.052	0.055
UN	2850° (at 2.5 atm N_2)	14.32	13.53	cubic, NaCl-type, $a = 4.890$ Å	0.0325	0.060
UP	2610°	10.22	9.05	cubic, NaCl-type, $a = 5.589$ Å	0.033	—
US	2480°	10.87	9.57	cubic, NaCl-type, $a = 5.485$ Å	0.021	—

It is obvious that the preparation of such phases as UC, or the mixed carbide (U, Pu)C call for accurate control of the composition in order to avoid the precipitation of metal or the separation of higher carbides. For this reason, phase diagram and thermodynamic studies of the uranium–carbon system have been necessary in order to ascertain the stability of a fuel under operational conditions. For instance, the successful design of the fuel pin requires that reaction between fuel and cladding shall be either very slow or absent.

Only molybdenum and tungsten are thermodynamically predicted to be truly compatible with UC. But the knowledge of rate processes is equally important. For instance, although the reaction:

$$UC + Nb \rightarrow U + NbC$$

with $\Delta G^0 = -14.5$ kcal/mole at $1000\,°C$, is thermodynamically possible, equilibrium is reached only at $1500\,°C$, and below $700\,°C$ reaction is not important. Hence, in addition to the measurement of thermodynamic data, thorough compatibility studies are required.

For economic reasons, the carbide fuel is fabricated from the oxide by a carbothermic reduction:

$$UO_2 + C \rightleftarrows UC + CO$$

This is followed by vacuum melting and casting. Basic studies have been made of the UO_2/C reaction between $1250°$ and $2000\,°C$, in support of this technical approach. In addition, fluidized bed processes, which employ either the carbothermic reduction and the reaction of UH_3 with a CH_4/H_2 mixture, are in the process of development.

But besides these chemical problems, a number of physical and metallurgical properties had to be examined in order satisfactorily to evaluate uranium monocarbide as a possible nuclear fuel. This is, for instance, especially true for the thermal conductivity of the material as a function of temperature and irradiation dose. Unfortunately, the measurements available show a large scatter; these arise from variations in carbon content which cause significant changes in the values. Parameters, such as swelling and fission gas release at high burn-up are equally important.

The properties of the semi-metallic compounds UN, UP, and US are considerably less well-known than UC, but they certainly justify a thorough consideration of these compounds as possible nuclear fuels[2,3]. Although, superficially, their properties are very similar, they appear to differ profoundly (Table 25), as a consequence of the increase in ionic character of the M–X bond which takes place as we move from UN to UP and then to US. Thus, the thermal conductivity of UC is approximately constant over the range $200–1250\,°C$, but that of uranium nitride rises markedly with increasing temperature (Fig. 39). Uranium monocarbide is easily hydrolyzed by water vapour, whereas UN and UP are both resistant to attack by steam above $100\,°C$. The compatibility of US with most metals is excellent, but its thermal conductivity and uranium density are lower than that of UC and UN.

A comparison of the properties of UN with UO_2 and UC (Table 25) shows that in some respects UN might be a superior nuclear fuel. But the rather high cross-section of N for thermal neutrons makes UN unsuitable as a nuclear fuel in thermal reactors. However, for applications in fast reactors, the properties of uranium mononitride are no doubt comparable with UC and, in some instances, even better. This is especially so when the simplicity and ease with which the stoichiometry of UN may be controlled is taken into consideration. Furthermore, it is much more stable towards water then UC and presents fewer compatibility problems with certain common cladding materials. For instance, the compatibility of UN, and incidentally of US, with stainless steel at 800 °C is much better than that of UC.

Irradiation experiments[4] confirm the good prospects which UN shows of being a reactor fuel, particularly in its favourable irradiation resistance at high burn-ups with fuel centre-line temperatures exceeding 1250 °C. Moreover, dimensional instability has not been observed, and fission gas release proved to be only 0.5 % of that produced in the fuel. Interestingly, the hydrogen formed by the ^{14}N (n, p)^{14}C reaction does not appear detrimental to the dimensionable stability of UN; dimensional and density measurements have not indicated any objectionable degree of swelling[4].

Uranium mononitride is, again for economic reasons, fabricated by the carbothermic reduction; in this nitrogen is passed over a mixture of oxide and carbon at 1400–1500 °C to give relatively pure mononitride:

$$UO_2 + C + \tfrac{1}{2} N_2 \leftrightarrows UN + CO$$

Unfortunately, the hypothetical UO shows an appreciable solubility in both UN and UC, and one obtains not the pure mononitride but a solid solution of the composition $UN_{1-x-y}C_xO_y$. Thermodynamic calculations have been made to define the single-phase region as functions of temperature and partial pressures of CO and N_2, as well as the maximum obtainable purity of

UN under different conditions. Similar calculations have been carried out for solid solutions $UN_{1-x}C_x$ giving the stability range and expected purities.

The product is a powder which can be densified to 93–98 % of the theoretical density by hot pressing at 1500–1700 °C under a modest pressure (3 ton/sq. inch). Hot pressing was also found to be the best method for preparing solid solutions of (U, Pu)N and (U, Pu)S.

Although the compatibility of UN with stainless steel is satisfactory, its compatibility with some other metals presents chemical problems. In general, it reacts thus:

$$UN + Zr \rightarrow ZrN + U,$$

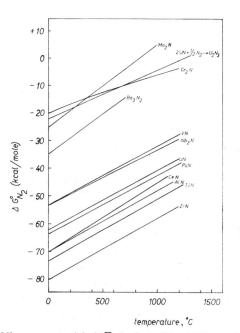

Fig. 39. Nitrogen potentials ($\Delta \bar{G}_{N_2}$) of possible cladding materials.

In this reaction, the nitrogen potential ($= \Delta \bar{G}_{N_2}$) determines whether chemical change shall take place or not, apart from

References p. 231

kinetic barriers[5]. Examples of nitrogen potentials are shown in Fig. 39 and illustrate that metals, such as aluminium, titanium or zirconium, are incompatible with UN.

The behaviour of the uranium compounds UN, UP and US at high temperatures is also different. The homogeneity range of UN widens out above 1500 °C and the lower phase boundary of $UN_{0.80}$ has been found at 2000 °C (see the phase diagram on p. 192). When they are heated in a vacuum above 1800 °C both UN and UP decompose, liberating liquid uranium. The greater stability of US is a desirable feature, especially when elements with very high centre temperatures ($>$ 2000 °C) are required. Moreover US is the only compound of this series which congruently sublimes at its true melting point.

Ternary solid solutions

In recent years, the ternary phases have proved to be of special interest for fuel applications since they possess better chemical and physical properties than the binary phases from which they are derived. The ternary phases in question, may be divided into two classes: (a) those with mixed cations, such as (U, Pu)N, and (b) those with mixed anions, such as U(C, N). The homogeneity range of these phases is principally determined by a difference in the lattice parameters of the components. The solid-solution behaviour of many of these systems, in particular the solubility limits they show, has been examined recently[6, 6a]. As could be expected, it appears that when the lattice parameters of the two phases differ by more than 10–15 %, which, for example, is the case with UN and UP, only a limited solubility occurs.

Solid solutions of U(C, N) are technologically very attractive, since they have a lower nitrogen dissociation pressure than pure uranium mononitride and a better stability towards water than uranium carbide. But their main advantage is that the solid solution can more easily be prepared from UO_2 than can the pure, binary phases. The conditions of temperature and pressure both of nitrogen and CO, required for the formation of the solid solution, according to:

$$UO_2 + (3 - x)C + \frac{x}{2}N_2 \rightleftarrows UC_{1-x}N_x + 2CO$$

have been examined in detail by Naoumidis and Stöcker[7]. They found that a stable, one-phase solid solution with values of x ranging from 0.1 to 0.6 could easily be prepared.

In spite of the simple structure and similarity of both components, many important details of the system are, however, not yet known; for instance, it is not known whether the solid solution is an ideal one or not. Other problems in relation to these solid solutions concern fabrication methods; the sintering processes which have been described, are technologically rather unattractive. The preparation of (U, Pu)N is not very easy by these means, since at high sintering temperatures, there is a tendency for decomposition to occur with the separation of plutonium metal. Hot-pressing at about 1600 °C is possible, and for the sulphides, as for the carbides, casting can be employed.

In conclusion, it may be said that in the promotion of their application as a nuclear fuel much more technological information is needed about these semi-metallic compounds. There is, for instance, a considerable lack of information on the behaviour of the fission products, formed during irradiation, in particular on their reaction with the parent nuclear fuel. Thus, thermodynamic information about the rare-earth carbides is lacking and is urgently needed. The carbide CeC is known to be more stable than UC, and its formation together with ternary carbides such as $UMoC_2$, in carbide-fuel ought to be considered. The same is true for the nitrogen formed during fission in uranium nitride fuel; here again, more thermodynamic information on the nitrides of several metals should be measured.

Reprocessing

During nuclear fission fuel consumption and the partial replacement of atoms of the fuel by neutron-absorbing fission products causes a fall in the reactivity of the effective loading. Hence, it becomes necessary to replace fuel elements in the reactor although they still contain most of their original fissionable material. The reprocessing of such spent fuel-elements presents an important problem in nuclear technology because of their high content of fissionable uranium, and the main methods employed will be discussed very briefly.

Aqueous reprocessing

Usually, the irradiated fuel is dissolved in aqueous solutions of nitric acid and solvent extraction is used for the separation of uranium and plutonium from each other and from the fission products. At present, all industrial methods still employ aqueous reprocessing.

Metallic fuel is readily soluble in 8–13 N HNO_3 and UO_2 pellets also dissolve in the same medium. In some procedures, such as the *Darex* process, the whole fuel element, including the cladding, is dissolved in an HNO_3/HCl mixture. In a variant of this process, the "chop and leach" method, the fuel element is cut into segments a few centimeters in length before being treated with nitric acid. However, all forms of chemical decanning lead to a high loss of uranium and plutonium (0.2–4.5 %) and are for this reason not favoured for the highly rated fuel-elements from fast reactors[8].

After the fuel has been brought into solution, the particular solvent extraction and subsequent processing method to be used naturally depends on the composition and enrichment which has been employed in preparing the original fuel.

In the *Redox* process for the processing of irradiated natural uranium, hexone is used as a solvent; and in the well-known *Purex* process the solvent is tributyl phosphate (TBP) dissolved in an inert hydrocarbon. Both processes use the redox reactions of plutonium to effect its separation from uranium. In the Redox process both uranium and plutonium(VI) in the organic phase, are separated together, in the first extraction-step, from the fission products present in the aqueous phase. Back-extraction of plutonium is done in the second extraction-step, in which the PuO_2^{2+} ion is reduced to the Pu^{3+} ion with ferrous sulphamate; by this means uranium remains in the organic phase and plutonium(III) is transferred to the aqueous phase. The most readily extractable oxidation state of plutonium in the Purex process is the $+4$ state, obtained by oxidation with sodium nitrite, rather than the $+6$ state. The separation of plutonium from uranium is done as in the Redox process by reduction of Pu(IV) to Pu(III) with ferrous sulphamate. In the final stages the uranium and plutonium which have been separated, are purified by solvent extraction with TBP.

The fluoride volatility process[9]

Since uranium is among the very few metallic elements that form highly volatile fluorides (see Chapter 9), it is possible to use very selective processes for the separation of uranium from the contaminants. The few volatile fluorides of the fission products, for instance, MoF_6, NbF_5, RuF_5 and TeF_6, which can be formed when irradiated uranium is fluorinated, differ sufficiently in volatility from UF_6 and PuF_6 and, in general, very high decontamination factors have been reported.

Since fluorine is a very reactive gas, the fluorination reaction is difficult to control and for this reason fluorination by liquids, particularly by ClF_3, has received serious attention. After the reaction:

$$U + 3 \, ClF_3 \rightarrow UF_6 + 3 \, ClF$$

the UF_6 is distilled off, nearly all the fission products being left in a residue consisting of solid fluorides. Since, however, some fission-product fluorides, such as those of niobium and ruthenium, volatilize appreciably, further decontamination of the UF_6 is achieved by its fractional distillation. The spent fluorinating reagent is regenerated by treating it with fluorine under pressure.

Selective fluorination of uranium to UF_6 in a fluid-bed reactor filled with alumina has been developed for the recovery of uranium and plutonium from UO_2 fuel; in this process ClF_3 is again used as the fluorinating reagent. Plutonium is recovered from the solid alumina in the bed as PuF_6 in a subsequent step by the action of fluorine at temperatures up to $550\,°C$[10]. The volatile fluorides passing from the reactor can be recovered either by condensation or by an absorption–desorption cycle on solid inorganic fluorides such as NaF:

$$UF_6 + 3\,NaF \xrightarrow{\;100\,°C\;} Na_3UF_9$$

When the temperature is raised to $400\,°C$, a stream of fluorine removes uranium, leaving the contaminants behind.

Several difficulties connected with the quantitive recovery of uranium and plutonium and with the instability of $PuF_6\,(\to PuF_4)$ have prevented this promising process from being applied on an industrial scale up to the present.

Pyrochemical processes

Other non-aqueous reprocessing methods are the pyrochemical processes which appear to show greatest promise in the recovery of fissionable material from fast reactor fuel. Some of these are based on physical changes which do not involve changes in the oxidation state of the fuel, whereas others rely on the selective oxidation of fission products.

An important example is the melt-refining process in the fuel cycle of the EBR-II. This process is based on the fact that at temperatures of $1200°$ to $1400\,°C$ in an argon atmosphere, fission products, such as barium, strontium, yttrium and the rare earths in the molten fuel (uranium$+5\,\%$ fissium), react with ceramic crucibles, of zirconia for instance, to form oxide slags, while other fission products, such as iodine and the gases xenon and krypton, volatilize. This process effects a substantial uranium purification, but has the disadvantage that the purified uranium must be refabricated by remote casting, since it is far from being completely decontaminated.

For reprocessing of ceramic fast-reactor fuel (oxide or carbide), a different pyrochemical process has been developed[11]. The fuel is suspended in a salt mixture $(MgCl_2,\ CaCl_2 + CaF_2)$ and is reduced by a Cu/Mg alloy at $800\,°C$. In this step the volatile fission products are released and the alkaline, alkaline-earth and rare-earth fission products remain in the salt phase. At the same time, uranium, plutonium and the noble-metal fission products are reduced to the metals and are then absorbed by the Cu/Mg alloy, from which uranium is precipitated while the plutonium remains in solution. The precipitated uranium is recovered by decanting the supernatant Cu/Mg alloy, washing with liquid magnesium to remove copper and the remaining fission products, and vacuum retorting[12] to separate residual magnesium from the metal. But it should be said that high decontamination factors cannot be reached in this process.

REFERENCES

1   T. L. MARKIN, V. J. WHEELER AND R. J. BONES, *J. Inorg. Nucl. Chem.*, 30 (1968) 807.
2   M. ALLBUTT AND R. M. DELL, *J. Nucl. Mat.*, 24 (1967) 1.
3   J. B. MOSER AND O. L. KRUGER, *J. Appl. Phys.*, 38 (1967) 3215.
4   R. A. WULLAERT, J. E. GATES AND J. BUGL, *Ceramic Bull.*, 43 (1964) 836.
5   M. ALLBUTT AND A.R. JUNKISON, *AERE-R-Report* 5466 (1967).
6   S. IMOTO, K. NIIHARA AND H. J. STÖCKER, *Thermodynamics of Nuclear Materials, Proceedings of a Symposium*, Vienna (1967), IAEA, Vienna (1968), p. 371.
6a Y. BASKIN, *Trans. Metall. Soc. AIME*, 239 (1967) 1708.
7   A. NAOUMIDIS AND H. J. STÖCKER, ref. 6, p. 287.
8   E. DETILLEUX, E. LOPEZ-MENCHERO, F. OSZUSKY, *Plutonium as a Reactor Fuel, Proceedings of a Symposium*, Brussels (*1967*), IAEA, Vienna (1967), p. 511.
9   R. C. VOGEL *et al.*, *Third International Conference on the Peaceful Uses of Atomic Energy*, Geneva (*1964*), United Nations (1965), Vol. 10, p. 491.
10 J. SCHMETS, G. CAMOZZO, R. HEREMANS AND G. PIERINI, *ANL-Trans.* 478 (1967).
11 *ANL-Report* 7349 (1967).
12 C. L. CHERNICK AND A. GLASSNER, *ANL-Report* 7399 (1967).

CHAPTER 14

Aspects of the Analytical Chemistry of Uranium

Qualitative identification of uranium

There are a number of reactions by which uranium may be identified qualitatively. A brown precipitate of UO_2S is obtained when ammonium sulphide is added to a solution containing uranyl ions. The precipitate is soluble in dilute mineral acids. Addition of diammonium phosphate to uranyl ions gives a pale-yellow precipitate of UO_2HPO_4 which is insoluble in acetic acid. With tri-ammonium phosphate the even less soluble compound $NH_4UO_2PO_4$ is precipitated.

The reaction with $K_4Fe(CN)_6$ is commonly used for the identification of uranyl ions. It gives a brown precipitate of $K_2UO_2Fe(CN)_6$, similar to the precipitates formed by copper and molybdenum. When, however, ammonia is added to the precipitate, it becomes yellow (without dissolution) in contrast with the precipitates produced by copper and molybdenum. The reaction may be carried out as a spot-test or on filter paper. In the latter a brown ring is formed, which may be distinguished from the brown rings of copper and molybdenum again by the reaction with ammonia.

Uranium is identified under the microscope as $TlUO_2(CO_3)_3$, which appears as pale-yellow rhombs when a small piece of solid $TlNO_3$ is added to a solution of a uranyl salt in ammonium carbonate. Although the yellow rhombs are identical with those given by thorium, they may be distinguished from the latter by the brown colour produced when $K_4Fe(CN)_6$ is added.

It should be emphasized that none of these reactions is specific, and that uranium is best identified spectrographically, the line at 4244.35 Å being used.

References p. 237

Quantitative determination of uranium

There are many methods for the quantitative determination of uranium: gravimetric, volumetric, coulometric, polarographic and others. Which method is to be used depends primarily on the nature of the sample being analyzed and on the amount of uranium in it. A comprehensive review of the subject has been written by Rodden[1]. Therefore, only a brief resumé of the main methods will be given here.

Gravimetric methods

The gravimetric determination of uranium is used only when macro-amounts of uranium are available. The two principal methods are the determination as U_3O_8 or by precipitation with organic reagents, of which oxine is usually used.

The determination as U_3O_8. The sample is ignited in air at about 850°C for several hours, and weighed in the form of the oxide U_3O_8; this oxide loses oxygen when heated above 800 °C, but takes up the oxygen rapidly again below that temperature to give the stoichiometric oxide. The method is very accurate for substances, such as oxides, nitrates and sulphates which decompose into U_3O_8. When, however, non-volatile components are present, they must first be separated. This is done by dissolving the sample in dilute acid, filtering off any solid insoluble in the acid, and precipitating the uranium as ammonium uranate (ADU) by adding aqueous ammonia. Soluble impurities must not be such as are precipitated by ammonia. The ADU is ignited to U_3O_8 which is weighed.

The determination with o-oxyquinoline (oxine). The oxine is added to the uranium solution which is buffered at a pH between 5 and 9. The red-brown precipitate, after drying at 105–110°C, contains 33.84 wt. % uranium and has the composition:

$$UO_2(C_9H_6NO)_2 \cdot C_9H_7NO$$

It has been found[2] that the pH during precipitation, the washing procedure, and the drying temperature are all critical for accurate determination.

Volumetric methods

The volumetric determination of uranium is based on the oxidation of U^{4+} ions with oxidizing agents, such as potassium dichromate or cerium(IV) sulphate solutions.

The uranium in solution is first reduced to U(IV) in a lead reductor and is then titrated, for instance with a standard potassium dichromate solution. Occasionally a zinc reductor is used, but with zinc a mixture of trivalent and tetravalent uranium is generally obtained. The trivalent uranium is usually oxidized to the tetravalent state by a current of air. In the case of the lead reductor, it appears that only tetravalent uranium is formed. Substances other than uranium that are reduced by lead and subsequently titrated must be absent from the original solution. Nitrates and nitrites are partially reduced and must be removed by fuming with sulphuric acid.

Therefore, when nitrate ions are present, the reduction of uranium with $TiCl_3$ is often preferred[3]. In this method an equivalent amount of Ti(IV) is produced and the excess of Ti(III) ions is oxidized by the nitrate ions. Sulphamic acid ($H_2N \cdot SO_3H$) is added to remove any HNO_2 that may be formed. The uranium is now directly titrated with a standard Ce(IV) solution, according to the equation:

$$U^{4+} + 2 H_2O + 2 Ce^{4+} \rightarrow UO_2^{2+} + 2 Ce^{3+} + 4 H^+$$

The end point of the titration may be observed potentiometrically.

The determination of the O/U ratio in uranium oxides.

This is required when the deviation from stoichiometry, for instance in UO_{2+x}, has to be determined. The sample is dissolved in warm, concentrated phosphoric acid and can be titrated directly, for instance with standard titanium(III) solution to give the uranium(VI) content[4]. For the determination of the O/U ratio in sintered uranium dioxide pellets, a sensitive method is required since the amount of uranium(VI), and consequently the value of x in UO_{2+x} is usually very small. Recently, Engelsman *et al.*[5] have reported such a method in which uranium(VI) is reduced by titration with iron(II) ammonium sulphate, after which the total amount of uranium is oxidized by titration with potassium dichromate.

The pellets are disssolved in concentrated phosphoric acid at 190°C and the end-points are detected electrometrically. The authors claim that a determination of x as low as 0.0001 is possible. Other methods for the determination of the O/U ratio in UO_{2+x} have been reviewed by Nickel et al.[6].

Other methods

The uranium content of solutions, containing about 25 mg uranium, can be found accurately by controlled-potential coulometry[7]. The method is also applicable for uranium–aluminium alloys containing about 24 % uranium. A standard deviation of 0.05 % on the uranium content is reported. The uranium is organic solutions, such as in tributyl phosphate, may be determined spectrophotometrically[8]. The method is applicable to small amounts of uranium, ranging from 1% down to a few thousands of a percent. The method is especially sensitive when pure uranium solutions are used.

Other methods for the determination of small amounts of uranium with rather high accuracies are polarography, colorimetry, and constant-current coulometry. If the available quantity of uranium is below 1 mg only refined techniques are used. Thus, amounts of uranium ranging from 1 mg down to 1 μg can be measured polarographically[9], but very small amounts, as little as 10^{-11} g of uranium, are best measured fluorimetrically. In this method uranium in a sodium fluoride flux is determined by its characteristic yellow-green fluorescence excited by ultraviolet radiation (3650 Å). Then uranyl ions give a specific fluorescence radiation at 5550 Å. The intensity of this radiation is proportional to the amount of uranium present in the range of 10^{-5} to 10^{-11} g uranium, and can be measured with a photoelectric meter.

Determination of impurities in uranium

The accurate and rapid determination of a large number of elements in the ppm or lower range is required for uranium used throughout the nuclear fuel cycle since, for this purpose it must not contain elements of high neutron absorption cross section.

Specifications of the amounts of impurities which can be tolerated have been given on p. 19. Most analyses for impurities have been performed by emission spectroscopy or colorimetry. In the first method, uranium is selectively removed, for instance by a single extraction with tributylphosphate (TBP) and the emission spectrum analyzed, or the uranium spectrum can be suppressed by a suitable carrier material and the use of a special form of the electrode[10]. More accurate is the colorimetric method, which is, however, time-consuming.

The determination of impurities by atomic absorption has been reported [11] to give a slightly better precision than colorimetry; the emission spectroscopic results tend to be low and much less precise than either of the other methods.

REFERENCES

1   C. J. RODDEN, *Analysis of Essential Nuclear Reactor Materials*, U. S. Government Printing Office, Washington D. C. (1964).
2   A. CLAASSEN AND J. VISSER, *Rec. Trav. Chim.*, 65 (1946) 211.
3   J. CORPEL AND F. REGNAUD, *Anal. Chim. Acta*, 27 (1962) 36.
4   J. R. SIMMLER, *NYO-Report* 5218 (1948).
5   J. J. ENGELSMAN, J. KNAAPE AND J. VISSER, *Talanta*, 15 (1968) 171.
6   H. NICKEL, *Nukleonik*, 8 (1966) 366.
7   Y. LE DUIGOU AND K. F. LAUER, *Euratom Report*, Eur 3601 f (1967).
8   Y. LE DUIGOU AND H. F. LAUER, *Anal. Chim. Acta*, 40 (1968) 534.
9   V. VERDINGH AND K. F. LAUER, *Z. Anal. Chem.*, 235 (1968) 311.
10 B. F. SCRIBNER AND H. R. MULLIN, *J. Res. Nat. Bur. Standards*, 37 (1946) 379.
11 C. R. WALKER AND O. A. VITA, *Anal. Chim. Acta*, 43 (1968) 27.

Subject Index